T0214283

Lecture Notes
in Business Information Processing 401

Series Editors

Wil van der Aalst ⓘ
RWTH Aachen University, Aachen, Germany
John Mylopoulos ⓘ
University of Trento, Trento, Italy
Michael Rosemann ⓘ
Queensland University of Technology, Brisbane, QLD, Australia
Michael J. Shaw
University of Illinois, Urbana-Champaign, IL, USA
Clemens Szyperski
Microsoft Research, Redmond, WA, USA

Benjamin Clapham · Jascha-Alexander Koch (Eds.)

Enterprise Applications, Markets and Services in the Finance Industry

10th International Workshop, FinanceCom 2020
Helsinki, Finland, August 18, 2020
Revised Selected Papers

 Springer

Editors
Benjamin Clapham ⓘ
Goethe University Frankfurt
Frankfurt, Germany

Jascha-Alexander Koch ⓘ
Goethe University Frankfurt
Frankfurt, Germany

ISSN 1865-1348   ISSN 1865-1356   (electronic)
Lecture Notes in Business Information Processing
ISBN 978-3-030-64465-9   ISBN 978-3-030-64466-6   (eBook)
https://doi.org/10.1007/978-3-030-64466-6

This Springer imprint is published by the registered company Springer Nature Switzerland AG
The registered company address is: Gewerbestrasse 11, 6330 Cham, Switzerland

Preface

Advancements in information technology not only changed the way we communicate and process data but particularly paved the way to new business models, markets, networks, services, and players in the financial services industry. Electronic trading, data analytics, and FinTech offerings represent only some of these developments. The FinanceCom workshop series has been providing significant academic research in this area at the intersection of information systems and finance from the beginning and well before the rise of the FinTech concept. Research presented at FinanceCom workshops aims to help academics and practitioners to understand, drive, and exploit the opportunities associated with these information technology-driven developments in the financial sector.

After the very successful FinanceCom workshops in Sydney, Australia (twice); Regensburg, Germany; Manchester, UK; Montreal, Canada; Paris, France; Frankfurt, Germany (twice); and Barcelona, Spain, FinanceCom 2020 was held virtually for the first time due to the global COVID-19 pandemic and associated travel restrictions. Thanks to our authors, presenters, and participants, FinanceCom 2020 led to fruitful discussions about the presented papers and an exchange of ideas despite the physical distance with participants joining the virtual conference room from Australia, Europe, and the USA.

For this first virtually held FinanceCom workshop, we received 14 submissions, of which we selected 6 high-quality papers to be presented and published after their revision in this volume together with an additional invited short paper reflecting the invited talk. The selection was based on a rigorous review process accomplished with the help of a Program Committee consisting of internationally renowned researchers in the field, who also significantly helped to improve the selected papers with their comments and suggestions.

This proceedings volume is structured in three comprehensible parts, each of which contains two thematically related papers. The first part contains contributions shedding light on machine learning applications in trading and financial markets. The first paper "State-of-the-Art in Applying Machine Learning to Electronic Trading" by Rabhi et al. presents a literature review providing insights on how machine learning techniques are being used for trading in electronic financial markets. In particular, the paper examines the target areas, the applied methods, and the purpose of machine learning applications for electronic trading. Thereby, Rabhi et al. identify gaps and opportunities for further research in this rapidly evolving field, which are especially related to information representation for machine learning-powered analytics, more sophisticated machine learning techniques, and enhanced automated trading strategies. Moreover, the authors call for multidisciplinary research approaches adopted with a strong industry focus to further advance this field. The second paper "Using Machine Learning to Predict Short-Term Movements of the Bitcoin Market" by Jaquart et al. applies a variety of machine learning models to test the predictability of the bitcoin market across different time horizons ranging from 1 to 60 minutes. Based on a comprehensive feature set, including technical, blockchain-based, sentiment-/interest-based, and asset-based features, the authors show that especially recurrent neural networks and gradient boosting classifiers are well-suited for predicting short-term bitcoin market movements outperforming a random classifier. Although a long-short trading strategy based on the classifier results generates monthly returns of up to 31% before transaction costs, it leads to negative returns after taking transaction costs into account.

The second part of the proceedings contains two contributions in the area of fraud detection and information generation in finance. The first paper "Scalable and Imbalance-Resistant Machine Learning Models for Anti-money Laundering: A Two-Layered Approach" by Tertychnyi et al. addresses the question of how to train accurate machine learning models supporting anti-money laundering provisions. In this particular use case, machine learning models need to fulfill the requirements of scalability and imbalance resistance to be valuable since they are applied to very large transaction datasets with high class imbalance due to the low number of instances indicating potentially illicit behavior. In order to achieve this goal, Tertychnyi et al. develop a two-layered approach consisting of a simple model in the first step and a more complex model in the second step, which is only applied to customers that could not be classified as non-illicit with high confidence in the first model. Besides the scalability of this approach, it also partially addresses the problem of high class imbalance since the first layer acts like an undersampling method for the second layer. In the other contribution of this section entitled "Leveraging Textual Analyst Sentiment for Investment," Palmer and Schäfer make use of natural language processing and analyze the informativeness of the sentiment conveyed in research reports written by financial analysts for both contemporaneous and future stock returns. The results show that a portfolio trading strategy exploiting textual sentiment in analyst reports generates an average monthly factor-adjusted return of 0.7 %. Consequently, the authors conclude that analysts provide valuable information for interpreting and predicting stock price movements in their research reports. These findings emphasize the importance of including qualitative information in prediction models in the area of financial markets.

The third part of the proceedings contains two papers focusing on alternative trading and investment offerings by FinTechs. In their paper "Portfolio Rankings on Social Trading Platforms in Uncertain Times," Bankamp and Muntermann analyze the value of portfolio rankings on social trading platforms regarding performance and especially regarding the successful management of market exposure and risk. For their analysis and in order to investigate the value of rankings as a protective function in turbulent market phases, the authors look at the stock market crash that occurred in connection with the COVID-19 pandemic in spring 2020. The findings show that portfolio rankings on social trading platforms indeed provide value for investors and protect them from extreme losses in downward periods. The second paper in this section, "What do Robo-Advisors Recommend? - An Analysis of Portfolio Structure, Performance and Risk," by Torno and Schildmann, investigates the quality of portfolio recommendations provided by robo-advisors that automatically recommend personalized portfolios based on customers' risk affinity and investment goals. Using 6 model customers obtaining portfolio recommendations from 36 different robo-advisors, the authors find that robo-advisors in fact provide distinct recommendations for different risk affinities and investment horizons, yet the recommended portfolios are less distinct for differences in investment horizons and consist of a high share of equities even for short-term investments.

The proceedings collection ends with a short paper summarizing the main points of the invited talk by Pradeep Kumar Ray on "The Financial Viability of eHealth and mHealth" in light of the ongoing COVID-19 pandemic. Prof. Ray and his coauthors elaborate on the benefits of eHealth (healthcare using information and communication technologies) and mHealth (healthcare using mobile phones). Since these technological solutions allow treating patients without face-to-face contact between patients and health professionals, they are particularly valuable in pandemic situations, and, thus, their use multiplied in the COVID-19 pandemic. Nevertheless, Prof. Ray and his coauthors argue that the economic viability of eHealth and mHealth is an extremely important aspect for their successful global adoption. Therefore, the authors derive a mathematical model to compare the costs associated with eHealth/mHealth with those of traditional face-to-face, paper-based methods. The application of the model shows cost savings in eHealth/mHealth but the exact amount depends on various factors such as the nature of healthcare or the country of application.

We would like to thank Fethi A. Rabhi for his invaluable guidance during the organization of the first virtual FinanceCom workshop and its proceedings. We are also grateful to our reviewers, authors, and the Program Committee members for the extraordinary work on the contents of this volume and to Ralf Gerstner and Christine Reiss from Springer for their excellent support in producing the FinanceCom 2020 proceedings.

October 2020

Benjamin Clapham
Jascha-Alexander Koch

Organization

Location

Due to the COVID-19 pandemic in 2020, the workshop was held virtually for the first time in its history.

Organizing Committee and Program Chairs

Benjamin Clapham   Goethe University Frankfurt, Germany
Jascha-Alexander Koch   Goethe University Frankfurt, Germany

Program Committee

Peter Gomber   Goethe University Frankfurt, Germany
Stefan Lessmann   Humboldt University of Berlin, Germany
Bernhard Lutz   Albert Ludwigs University of Freiburg, Germany
Nikolay Mehandjiev   The University of Manchester, UK
Jan Muntermann   University of Göttingen, Germany
Dirk Neumann   University of Freiburg, Germany
Nicolas Pröllochs   Justus Liebig University of Giessen, Germany
Fethi A. Rabhi   University of New South Wales, Australia
Brahim Saadouni   The University of Manchester, UK
Michael Siering   Goethe University Frankfurt, Germany
Andrea Signori   Università Cattolica del Sacro Cuore, Italy
Basem Suleiman   The University of Sydney, Australia
Axel Winkelmann   University of Würzburg, Germany

Steering Committee for the FinanceCom-Workshop Series

Peter Gomber   Goethe University Frankfurt, Germany
Dennis Kundisch   Paderborn University, Germany
Nikolay Mehandjiev   The University of Manchester, UK
Jan Muntermann   University of Göttingen, Germany
Dirk Neumann   University of Freiburg, Germany
Fethi A. Rabhi   University of New South Wales, Australia
Federico Rajola   Università Cattolica del Sacro Cuore, Italy
Ryan Riordan   Smith School of Business at Queen's University, Canada
Christof Weinhardt   Karlsruhe Institute of Technology, Germany

Contents

Machine Learning Applications in Trading and Financial Markets

State-of-the-Art in Applying Machine Learning to Electronic Trading

Fethi A. Rabhi[1]([✉]), Nikolay Mehandjiev[2], and Ali Baghdadi[3]

[1] School of Computer Science and Engineering, University of New South Wales,
Sydney, Australia
f.rabhi@unsw.edu.au
[2] Alliance Manchester Business School, The University of Manchester, Manchester, UK
n.mehandjiev@manchester.ac.uk
[3] Researcher and Consultant, 3814974695 Arak, Iran
Alibaghdadi9574@gmail.com

Abstract. This paper presents a literature survey of how machine learning techniques are being used in the area of electronic financial market trading. It first defines the essential components of an electronic trading system. It then examines some existing research efforts in applying machine learning techniques to the area of electronic trading, examining the target areas, methods used and their purpose. It also identifies the gaps and opportunities for further research in this new expanding field.

Keywords: Machine learning · Electronic trading · Financial markets · Algorithmic trading

1 Introduction

The area of electronic trading has seen dramatic changes in recent years. In particular, trading in financial markets has become a global activity because of the recent technological developments that have facilitated the instantaneous exchange of information, securities and funds worldwide. The economic factors behind this push are transparency, cost, risk management and the potential for anonymity. Electronic trading not only improves transparency of prevailing prices in a market but also provides more information such as the depth of a market which indicates the potential supply and demand away from the current market price. Taking advantage of economies of scale, the cost of trading becomes low as investors can gather information quickly, forcing greater competition among market participants. The use of 'straight-through processing' not only reduces costs of settling transactions but also offers better risk management to market participants as it minimises the risks of fraud and human errors associated with manual trade processing [41].

Given the huge variety in exchanges and multilateral trading venues, many market participants now employ algorithmic trading to automatically make certain trading decisions, submit orders and manage those orders after submission [51]. Advanced market

B. Clapham and J.-A. Koch (Eds.): FinanceCom 2020, LNBIP 401, pp. 3–20, 2020.
https://doi.org/10.1007/978-3-030-64466-6_1

data feeds as well the integration of news and social media information sources now allow for the full observation of market participants' actions and the impact of these actions. Increasingly, more sophisticated approaches to automated execution are being deployed in which different methods and different algorithms could be used to fit the constraints of different classes of transactions and under differing market circumstances.

At the same time, the exponential increase in computing power and data storage during the last decade has resulted in the rapid development of machine learning and data mining with diverse applications in economics, finance, science, engineering, and technology. In the finance area, machine learning models have elicited considerable attention from many researchers because of their predictive power [27]. Compared to statistically rigorous analytic methods, machine learning techniques provide users with more control over the accuracy/performance trade-off: at one extreme giving exact results (but requiring huge computational power to process vast amounts of data) and at the other extreme providing rough estimations (in a computationally cheap and hence fast way). From "robo-advisers" to high frequency trading, the market has become ever-more dynamic, efficient, and competitive [20]. In particular, the use of machine learning techniques has the potential to drastically change the ways trading is conducted, e.g. applying smart order routing or intelligently determining order size. However, their uptake depends on a complex interplay of technology, personal and organizational factors, and the complex nature of machine learning techniques makes its outcomes less transparent. This impedes the building of trust in the technology.

Initiated by econometricians, statisticians and computer scientists, the interest of the academic community for machine learning techniques is growing but there is still a large gap with professional practice [25]. One of the reasons is that different research communities understand machine learning in different ways. To make matters worse, this concept is also used interchangeably with that of Artificial Intelligence or AI. In this paper, we use the term "machine learning techniques" as understood by practitioners. This is illustrated by a recent survey conducted by the Bank of England and the Financial Conduct Authority in October 2019 [4] which defines machine learning as "a methodology in which programs fit a model or recognize patterns from data, without explicitly programmed and with limited or no human intervention". They consider that machine learning constitutes an improvement rather than a fundamental change from statistical methods. Machine learning can also work hand in hand with any type of automated optimization techniques that allow a computer to "twiddle the knobs", i.e. to search the space of possible combinations of parameter values, to find good settings [7]. For these reasons, we use a fairly broad definition of machine learning that includes advanced statistical techniques and any data intensive method that extracts patterns from data including Artificial Neural Networks (ANN), Agent-based approaches [45] and evolutionary computing [23].

The rest of this paper discusses recent research efforts aimed at applying machine learning techniques in the area of electronic trading. Section 2 defines the area in more detail and presents the different elements of an electronic trading system. Section 3 describes the methodology used in the survey. Section 4 presents the different papers according to their contributions in applying machine learning techniques within an electronic trading system. Section 5 concludes the paper.

2 Background

2.1 Components of an Electronic Trading System

In the context of this paper, electronic trading is associated with the area of financial markets so it is mainly concerned with the automation of traditional trading activities in the finance domain and their associated processes. This study will primarily concern order-driven (auction) markets in which trading occurs in different ways such as single-priced auction, continuous rule-based two-sided auction and crossing networks. In this type of markets, trader orders for any particular security are centralized in a single order book where buyers are seeking the lowest price and sellers are seeking the highest price. An order-driven market uses order precedence rules to match a buyer to sellers and trade pricing rules to price the resulting trades. As trading has become increasingly fragmented, this study will also consider the situation in which multiple order books are competing against each other, with variation in rules and pricing across the different books. In these cases, integration is achieved through smart order routers.

In such markets, trading decisions have also become more complex as they involve the cooperation of different participants and their systems interacting across wide geographical boundaries and different time zones. For these reasons, the use of machine learning techniques is being considered in different parts of the electronic trading life-cycle to increase efficiency and accuracy of such decisions. Many researchers believe that machine learning has the potential to replace theoretically derived models with data driven models [12].

We view electronic trading as a number of stages, each of which can be highly automated. As the scope of the study is related to the use of algorithms (and in particular machine learning) in this area, this encompasses the area of **algorithmic trading** which has generated abundant literature recently (e.g. books such as [22, 34, 37]).

This paper defines an electronic trading system as a system in which automation is applied wholly or partly in the following cycle (see Fig. 1):

- Upstream decision-making: these are decisions made before trading takes place such as defining the purpose behind trading or the trading strategy being used [39];
- Trading desks: their main role is in processing a wide range of information sources that will inform when and how trade will be conducted followed by how to manage order execution;
- Markets: provide the venues on which trade execution take place;
- Post-trade analytics: their role is to analyse market information for different purposes such as surveillance.

The exchange of information and processing of transactions can occur at a high or low frequency and the asset holding periods can be a few seconds, hours, days, weeks or months. In the rest of this section, we describe the different types of algorithms that play a role in this cycle.

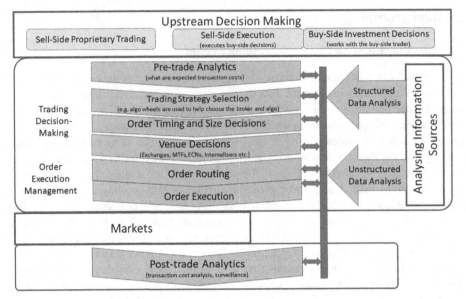

Fig. 1. Structure of trading activities

2.2 Upstream Decision Making

There are different categories of decision making that need to be made before trading: investment trading related decisions and proprietary trading decisions.

- Agent trading (trading for the account of the customer). This includes:

  – Investment (Buy side): these are decisions made on behalf of investors to satisfy some objectives such as transforming investment decisions into orders.
  – Sell-Side Execution: their role is to execute the sell-side of buy-side investment decisions.

- Proprietary trading: proprietary trading occurs when firms or individuals trade with their own money rather than service the needs of clients. This is also known as principal execution (trading on the broker's own account). According to [21], there are two types of proprietary trading:

  – Market making: the goal is not to build permanent positions but seek profits from market movements. The main decision in market making is the simultaneous generation of buy/sell quotes.
  – Statistical arbitrage: trade on information about relative asset values (opportunities can be short or long term).

In all these types decision-making, we have the same activities which are generating buy or sell orders but the goals are different.

2.3 Analysing Information Sources

Analysing information sources is a critical activity that feeds into several parts of an electronic trading system depending on the nature of the information. Generally, information sources are divided into structured and unstructured data.

Structured market information includes real-time trading data and events such as orders, trades, quotes, indices, and announcements [9]. It also includes historical information such as past trading data and companies' periodic reports. Additional statistical data and analytics may also be available; however, participants such as brokers can generate their own analytics based on raw historic market data. Market information dissemination is crucial to the fairness and efficiency of a market and hence contributes to its attractiveness to traders and encourages companies to be listed on the market.

In addition to structured data, unstructured data has also been used to generate signals related to various market conditions, and in particular news but more recently social media data.

The majority of stock market data analysis can be divided into three main sections: fundamental, technical, and sentiment analysis.

Fundamental analysis can be applied to both structured and unstructured data. A number of machine learning techniques have been widely used for financial classification problems, for example ANNs and Support Vector Machines (SVM) have been used to classify different fundamental variables for analysing structured data [16].

Technical analysis is mostly used to analyse structured data. Some approaches used machine learning besides technical analysis. For instance, Dash [13] explained a hybrid stock trading framework by integrating technical analysis with machine learning techniques.

Sentiment analysis is the most popular analysis for unstructured data, aiming to extract information about the underlying behavioural intentions of actors based on the sentiments expressed in the unstructured data produced by them. For example, Twitter sentiment analysis has been used as an antecedent of stock price trends [36].

In [42], the authors provide an overview of traditional news outlets that have been associated with the area of financial trading:

- **Macroeconomic announcements:** this is information regularly released by Governments regularly to inform the public regarding various features of economy such as the Gross Domestic Product, the Current Account Balance, the Consumer Price Index, and Unemployment statistics. These announcements are scheduled, so investors are aware when this information will be released.
- **Regulatory announcements:** this is information released to investors by companies listed on stock exchanges periodically e.g. earnings announcements. Typically, these announcements are in a fixed format to help investors compare information from multiple companies.
- **Press announcements:** issued by companies whenever they see fit e.g. when they wish to inform investors of a new product which has the potential to increase their profits, or to directly address rumours or issues raised by the press or members of the public. As the announcement is issued directly by the company which it affects, it can have more impact on its stock price than other non-regulatory news categories.

- **Analyst recommendations:** originating from analysts investigating all available information for given companies and recommending whether investors should buy, sell, or hold a given stock. Furthermore, analysts will often provide their clients with an in-depth analysis of the earnings potential for a given company.
- **Journalistic commentary:** any important news which does not fit into the above categories and generated by journalists unaffiliated with the company in question.

2.4   Trading Decision Making and Order Execution Management

We distinguish between decisions made during trading decision making and those made during order execution. *Trading decision-making* involves:

Trading Strategy Selection: these are decisions related to selecting the trading strategy to be used. Increasingly, *algo wheels* have become popular as a method for choosing a strategy. According to [2], "An algo wheel is an automated routing process which assigns a broker algo to orders from a pre-configured list of algo solutions. 'Algo wheel' is a broad term, encompassing fully automated solutions to mostly trader-directed flow".

Order timing and size decisions: these are related to choosing the size of an order and when it should be submitted to the venue.

Venue decisions: as trading desks can be connected to multiple brokers and liquidity sources, these are decisions related to the choice of a venue for executing the orders. There are many different types of venues besides traditional exchanges such as Multilateral Trading Facilities (MTFs) and Electronic Communications Networks (ECNs). Venues can require complete disclosure of counterparty details or allow full anonymity or provide many other options between these two extremes. Trades can also be executed internally within a large broker system for example. These decisions can be made statically or dynamically. Smart Order Routing refers to systems which direct orders to the best liquidity source. A broker can fill orders using different venues in the following manner [26]:

Order to the Floor: In some exchanges like New York stock exchange, the broker can direct an order to the floor of the stock exchange, or a regional exchange. In this way, because this order is going through human hands, it may cause some costs and take some time to fill.

Order to Third Market Maker: A Third Market Maker will receive the order in two cases: 1) They could be paying a commission to the broker to direct the order to them. 2) The broker is not a member firm of the best venue for the order.

Internalization: This happens when a broker decides to fill an order from its own inventory of stocks and make some money on the spread. If this route is selected, the order execution will be quite fast.

Electronic Communications Network (ECN): This method is especially suitable for limit orders because of the speed with which ECNs can match the prices.

Order to Market Maker: In this way, the broker can gain speed and additional commission by directing a trade to the market maker in charge of the specific stock.

Order execution management involves:

Order routing: sending the orders to the desired market venues. A trader has to monitor in real time every liquidity source and price levels for each venue.
Order execution: follow-up of the order through time and across markets and amending, deleting or changing the status of an order depending on market conditions.

Algorithms that support trading related decisions can be classified according to their objective: match given benchmarks (impact-driven algorithms), minimize transaction costs (cost-driven algorithms) and seek liquidity in different markets (smart order-routing algorithms). For example, a volume participation method uses predictions of future volume from a statistical model of past trading activity in the instrument being traded to decide when to place orders. A Volume-Weighted Average Price (VWAP) algorithm will use volume-participation methods to guarantee that a large stock trade will be decomposed into smaller orders that are submitted into the market in a way that the average price per share for the overall block trade will match the VWAP of the stock over the window for the trade.

In addition, there are many algorithms that play a role in reducing the latency in processing information and taking actions. These algorithms operate through the entire trading process, known as Straight-Through Processing (STP), from initiation to payments and clearing, is one seamless electronic flow of transaction-processing steps. A related process known as Direct Market Access (DMA) is where buy side investors are given direct access to the electronic order-books of an exchange, rather than having to interact with the market via a sell side intermediary or a broker. There are now many different technologies and ways of accessing market sources and submitting transactions, in many cases ranging from direct access to using various intermediaries to the market. According to [7], DMA and STP have now been joined by another third three-letter acronym: SOR, for Smart Order Routing.

2.5 Pre-trade and Post-trade Analytics

There are many ways of conducting predictive analysis which take place according to any type of model. Model types listed in [7] include Alpha models (to determine the forecasted value in an asset), Risk models, Transaction Cost models, Portfolio Construction models and Execution models.

For example, *Transaction Cost Analysis (TCA)* models help investors to reduce their trading costs, to manage their portfolio more accurately, and to improve the quality of execution of their trades. TCA achieves this by providing feedback to traders and providing greater transparency into their strategies. TCA can be used in Pre-trade and Post-trade analytics. Pre-trade TCA uses trade costs and market impact information from historical trades to make cost predictions about an order. In Post-trade TCA, traders can assess the quality of execution retrospectively, by comparing actual prices of execution to benchmark prices (e.g. VWAP for the trade within a narrow time period).

3 Survey Methodology

Having defined the scope of interest, we used the concepts developed and used in the scope to define a query string which we submitted to the Web of Science on-line database (v5.34) on 28th December 2019. The search contained three main parts (Technology, Field of study and Activity) which were joined by "AND" operator. The terms comprising each part are enumerated below and these were joined by "OR" operator.

Technology	Field	Activity
"artificial intelligence" | "machine learning" | AI | ML | "genetic algorithms" | "GA"	Financ*	trading | "order-driven" | "order execution" | "Order routing" | "order size" | "order execution" | "market analysis" | "sentiment analysis" | "fundamental analysis" | "technical analysis" | "market making"

The search results were limited to those in English and to the following document types: Article, Abstract of Published Item, Book, Book Chapter, Book Review, Editorial Material, Proceedings Paper.

The initial search produced 511 records, which were then pruned based on the area ascribed to the record in the Web of Science. We excluded records from the following areas as these were judged irrelevant to our search:

information science library science \| health policy services \| engineering manufacturing \| public environmental occupational health \| biochemistry molecular biology \| engineering biomedical \| chemistry analytical \| astronomy astrophysics \| engineering chemical \| chemistry applied \| engineering environmental \| chemistry multidisciplinary \| geography \| mathematics \| law \| mechanics \| hospitality leisure sport tourism \| physics applied \| nutrition dietetics \| biotechnology applied microbiology \| thermodynamics \| horticulture \| transportation science technology \| urban studies \| public administration \| health care sciences services \| physics multidisciplinary \| mathematical computational biology \| computer science hardware architecture \| neurosciences \| environmental studies \| ecology \| sociology \| environmental sciences \| polymer science \| zoology \| water resources \| agricultural engineering \| automation control systems \| agronomy \| genetics heredity \| political science \| history of social sciences \| construction building technology \| immunology \| development studies \| medical informatics \| engineering mechanical \| microbiology \| nanoscience nanotechnology \| international relations \| meteorology atmospheric sciences \| toxicology \| plant sciences \| veterinary sciences \| energy fuels \| agricultural economics policy \| education scientific disciplines \| telecommunications \| biochemical research methods \| chemistry physical \| infectious diseases \| materials science multidisciplinary \| forestry \| pharmacology pharmacy \| regional urban planning \| physics mathematical \| green sustainable science technology \| agriculture multidisciplinary \| computer science cybernetics \| robotics \| demography \| food science technology \| evolutionary biology \| engineering civil).

This resulted in 392 records which were subjected to further analysis below.

4 Results Per Categories

All research papers that have been identified in the search results have been categorised according to the following areas:

- Analysing data patterns: e.g. predicting market trends using machine learning techniques;
- Information processing and analysis: acquiring, pre-processing and analysing structured and unstructured data for the purpose of applying machine learning techniques;
- Trading decision making and order execution management: the use of machine learning for deciding the trading strategy and for managing the order lifecycle.

4.1 Analysing Data Patterns

This represents the largest body of research that was uncovered during this study because identifying patterns such as stock price movements has been a major challenge for both academics and practitioners for decades. It has been long known that markets are highly dynamic, complex, nonlinear, evolutionary and chaotic by nature [43]; partially because of their sensitivity to political factors, microeconomic and macroeconomic conditions, and investors' expectations. Therefore, prediction tasks are extremely difficult although many studies point out that markets in some cases exhibit a degree of regularity [35] which may help predictive modelling. Existing models that establish a relationship between return predictive signals, (the features) and future returns (the targets), are not able to capture complex non-linear dependencies [18].

To address the limitations of traditional econometric models, machine learning models have become popular for their ability to process complex, imprecise, and large amounts of data [35]. In addition, these methods enable the use of different types of data (qualitative and quantitative) and are not subject to rigid assumptions such as those imposed on econometric models. They have the freedom to incorporate fundamental and technical analysis into a forecasting model and can be adapted to different market conditions. A history and related research work related to the use of machine learning techniques for predicting stock prices is given in [43].

Examples of work in this area include [10] which uses supervised learning techniques for predicting the trend of the price in a single stock. Authors have implemented different supervised learning models and they have found that the SVM model can provide the most accurate prediction with 79% accuracy. Boonpeng and Jeatrakul [5] have applied the multi-binary classification using One-Against-One (OAO) and One-Against-All (OAA) techniques instead of usual machine learning methods like the SVM. In their paper, they show that the OAA technique outperforms other techniques. Lv *et al.* [33] have applied 12 machine learning algorithms with 44 technical indicators as input features to predict the trend of stock price. Also, they have divided stocks into 9 different industries and execute machine learning algorithms in each of these industries. Moreover, they have evaluated the performance of these algorithms in every industry and used the result to set some rules governing the selection of an optimal method for each industry.

Most studies tend to develop forecasting models for equity pricing using low frequency data (e.g. daily). More recent work has been applying machine learning for analysing intraday patterns [35] or for pricing financial derivatives [27]. More sophisticated machine learning techniques are also being developed to deal with the characteristics of high frequency data. For example, the model proposed in [18] applies long short-term memory (LSTM) networks, one of the deep learning techniques, for sequence learning tasks like time series prediction.

Evidence from these studies suggest that in general, machine learning techniques are able to identify (non-linear) structures in some very specific situations. Therefore, banks and financial institutions are investing heavily in the development of machine learning models and have started to deploy them in the financial trading arena [43]. However, these models have their limitations owing to the tremendous noise and complex dimensionality of stock price data, and besides, the quantity of data itself and the input variables may also interfere with each other.

Besides predictive analytics on asset prices, machine learning techniques have also been used in solving categorization problems. For example, [51] addresses the problem of categorizing and recognizing traders (or, equivalently, trading algorithms) on the basis of observed limit orders using a process from machine learning known as inverse reinforcement learning (IRL).

The work presented in this section is by no means exhaustive and intended to show the diversity of approaches used. Besides the use of conventional machine learning techniques, the survey by [23] describe extensive work in using evolutionary computation techniques in the area of technical analysis.

4.2 Information Processing and Analysis

The rise in the use of machine learning techniques has also triggered a lot of interest in developing innovative ways of gathering and processing data from multiple sources in order to produce good quality datasets. The motivation behind these efforts is that despite the availability of large amounts of datasets, these datasets are characterized by their massive size and high dimensionality and in need of further processing before they can be used. Related work differentiates between the analysis of structured and unstructured datasets as they present different characteristics.

An extensive survey of this area is presented in [24]. They classify structured data into three main types. The first one is raw data which may include fine-grained data or less-detailed data. The second type of data consists of diverse technical indicators calculated from raw data. The third type consists of economic indicators. These indicators can be prices or indices which are representative of the entire economic environment like gold price.

It is recognized that the direct use of low-level market data in any type of model is not recommended because of heterogeneity, noise accumulation, spurious correlations, and incidental endogeneity [17, 43]. Despite the fact that machine learning techniques can deal with more features than traditional econometrics models, adding more predictors is not a guarantee for increasing the performance of the analysis [25].

There has been extensive research work done in textual information representations that enable the exploration of relationships between news content and financial markets. For example, Das [11] defines the problem and outlines difficulties of analysing text in such a context. Amongst many examples of work in this area, Li [31] claim they could improve the accuracy of price prediction in the stock markets by integrating both market news and stock prices. Schumaker and Chen [46] propose a stock market forecasting engine based on financial news articles by a text classification approach. They show that their system outperforms some of the other systems.

More recent work in textual analysis has focused on social media data. For example, Porshnev, Redkin, and Shevchenko [38] survey how the analyses of tweets increase the accuracy of prediction for stock market indicators. There are many other examples of studies that study relationships between tweets and stock price movements [28, 36]. The survey by Hu et al. [23] describes some work in using evolutionary computation techniques in the area of analysing social media data with limited success. In general, any research work in the area of sentiment analysis is also relevant to the area of financial trading as sentiment affects markets. For example, Day and Lee [14] use deep learning to extract sentiments from a number of news outlets and Dhas, Vigila, and Star [15] report on having used big data to achieve a more accurate prediction by combining sentiment analysis and technical analysis.

The majority of known analysis methods use data representations that are based on physical time change or snapshots of the market, taken at fixed intervals. An alternative data modelling approach called Directional Changes (DC) records the key events in the market (e.g., changes in the stock price by a pre-specified percentage) and summarise the data based on these events, moving away from a physical-time view to an event-based-time view. Kampouridis and Otero [30] report that transforming low level data into a high level representation improves the accuracy of the predictions.

Again, there are numerous examples of research papers in this space and the ones above have been provided for illustrating the work in this area.

4.3 Trading Decision Making and Order Execution Management

Decision-making has become very complex at trading desks and the use of algorithms has been seen as essential to determine impact on market liquidity and volatility [21]. This reference also explains the importance of implementing automated safeguard mechanisms in order to ensure safe, fair, and orderly trading. The area of trading strategy selection and execution has seen a huge push towards automation due to pressure from regulators (e.g. best execution requirements). Hence, there is the potential to use machine learning techniques at different levels in this area.

Many challenges and related work associated with building (profitable) automatic trading algorithms with big data are discussed in [25]. Machine learning techniques have also been used to help solve complex optimization problems such as forming pairs of multiple assets under multiple objectives. For example, Goldkamp and Dehghanimo-hammadabadi [20] describe an approach for pairs trading based on genetic algorithms. There are few examples of research work in the area of order routing. In [8], the authors have formulated the optimal order placement problem for a market participant who can submit market orders and limit orders across various exchanges. They have used a supervised learning algorithm in their research and showed that a simultaneous placement of limit orders on multiple trading venues can cause a significant reduction of transaction costs.

The usage of machine learning has improved the robustness of automatic trading systems particularly in the High Frequency Trading (HFT) sector. Simple approaches include [1], which perform single-position, intraday automated trading based on a neuro-genetic algorithm and in which an artificial neural network provides trading signals to an automated trading agent. The system proposed in [6] performs risk management while opening and closing trading positions by using a performance-weighted ensemble of random forests to predict the expected profit of a trade given the prevailing market conditions. The model in [48] uses a hybrid method based on a popular ANN trend prediction model to dynamically prioritize higher performing regions at an intraday level and adapts money management policies with the objective to maximize global risk-adjusted performance. The authors in [44] propose an automatic trading system which is able to perform careful stock price prediction which is further improved by using adaptive correction based on the hypothesis that stock price formation is regulated by the Markov stochastic property. Other examples of work in this space include [45] who propose a technique in which trading agents learn and adapt their strategy by continuously analysing the data available.

In addition to the previous categories, there are examples of using machine learning to deal with operational efficiency problems. For instance, Huang et al. [24] bring some examples of using algorithms that are inspired by biological evolution. They state that these algorithms have shown great efficacy in discovering trading rules and optimizing global parameters. According to this paper, optimization in order execution is a strategy by which a system can execute a large number of orders, which should trade within a certain period of time, and with minimum expenditure. Li et al. [32] explain that since the speed of incoming data in HFT is at millisecond level, thus, in order to adapt to this condition and give sufficient time for the trading strategies to respond to the orders, the prediction speed of the system should be faster than speed of data. Therefore, they have designed a trading signal mining platform for sake of high prediction accuracy and fast prediction speed simultaneously using extreme learning machine (ELM), as a supervised technique to solve the speed problem. The ELM can randomly assign the input weights and hidden layer biases.

Despite these examples, academic research work is often lagging behind industrial innovations in this area, as evidenced by the proliferation of information about approaches and applications still coming from industry outlets. For example, some vendors (e.g. InfoReach TMS Algo Wheel [19]) are proposing products backed by machine learning algorithms to assist in trade strategy selection and best execution reporting. Psomadelis [40] describes an approach used in a Schroder's product called QTRMASTER. The algo wheel sits as a strategy within QTRMASTER and is configured to randomly select between a number of trading strategies. The traders at the different desks (typically region specialists) are responsible for selecting new algorithms to bring to the algo wheel. Execution data related to transaction costs is collected whenever a particular strategy is used, together with a wide range of variables and features associated with the corresponding orders. All execution data is grouped in a centralised repository which is used as a training set by a machine learning technique (a Random Forest Regression model) that makes recommendations for traders regarding the best strategy (i.e. one that minimises transaction costs) to use for a trade. A trader can choose a strategy as per the recommendation or reject it if they have data or information that was not taken into account previously. This may lead to a new algorithm being added to the algo wheel. Many other companies also plan to develop algo wheel products in-house [47].

5 Summary, Conclusions and Further Work

The purpose of this literature survey is to perform a preliminary analysis of the area of electronic trading and chart the landscape regarding how machine learning techniques are being used to improve the efficiency of financial markets trading. Unlike other surveys, it takes an industry-wide view when considering the different types of machine learning techniques currently being used, ensuring systematic coverage of all trading-related activities from a practical perspective.

The structure of the trading cycle, earlier presented in Fig. 1, is an additional contribution of this paper to the literature that can be used as a foundation of future studies and state-of-art analysis research. It illustrates some of the major trends that have been shaking this sector in recent years:

- **De-verticalization and move towards a service economy**: important changes in the regulatory frameworks (particularly the MiFID regulations in Europe) have disrupted the vertical separation between "sell side" (e.g. investment banking that had privileged access to market liquidity on centralized exchanges) and "buy-side" fund managers [7]. More stringent compliance requirements have changed the relationships between equity trading businesses and their brokers. Buy-side companies can now pick-and-choose their technology components as well as their trading venues and therefore avoid paying for the functionality previously provided by investment banks [49]. Consequently, the Fintech sector has witnessed a rise in Small and Medium-sized Enterprises (SMEs) offering software services and component technologies that can perform various functions in the trading cycle.
- **New alternative trading venues:** there has been significant growth in the provision of new alternatives to existing trading venues. This includes "dark pools" provided by off-exchange trading venues, called Alternative Trading Systems (ATSs) in the US and Multilateral Trading Facilities (MTFs) in Europe. They allow large blocks of shares to be traded with a higher degree of anonymity. Despite of the global increase in trading venues, 33% of traders in a recent survey still mention liquidity availability as an important issue in 2020 [29].
- **Extending information reach:** there has been a huge rise in the availability, variety and volume of data associated with the different models used during the trading cycle, as well as a diversification in the types of instruments that can be traded. In particular, complex derivatives and structured instruments such as exchange traded funds (ETFs) require additional information to be processed to enable complex predictions to be made. At the same time, the increasing use of sophisticated machine learning techniques (particularly deep learning) in pre-trade and post-trade analytics require more information to be processed to enable faster and more accurate decision-making. A recent survey is listing "analysis of previously inaccessible data" which is "reshaping trading strategies" as a key driver in this area [29].

- **Faster speed and more automation:** the trend in the use of high-speed, high band-width, adaptive technologies in automated systems interacting with a wide range of trading venues is set to continue over the next decade. Although high-frequency traders only represent a small proportion of trading parties, they are responsible for a majority of the equity trading volume. As most existing models and systems have been designed to operate with static data, there is now a need for new cost-effective solutions that are able to acquire, process and analyse real-time data. In addition, the need to integrate several systems (possibly from different vendors) has revealed many issues related to workflow design and efficiency. Finally, the tendency to use machine learning techniques to assist in selecting the most appropriate trading strategy based on a user's parameters is set to continue as only 27% of firms to date have tried "AI-powered algos" according to a recent survey [29].
- **Need for better risk management:** all of the above trends and in particular the need to integrate multiple technologies and models exacerbate the potential for massive risk throughout the trading cycle. In addition, the risks associated with automated trading are poorly understood especially the quantification of risk which requires a new risk management culture to be established in financial institutions and regulators [7].

In general, the survey reveals a mismatch between the focal areas of academic researchers, which generally reflect classic academic disciplines divisions. For example, most studies in computer science tend to focus on the computerized analysis of information (both structured and unstructured) for the purpose of forecasting stock prices. There is also a wide gap between academic finance and "professional" finance when it comes to analysing big datasets with machine learning methods and putting these models into operation within a realistic trading environment.

Future research work in the area of applying machine learning techniques to electronic trading will require multidisciplinary approaches to be adopted with a strong industry focus. We notably identify the following important research topics that are defined around the challenges outlined earlier:

- **Information representation for machine-learning powered analytics:** feeding hundreds of low-level indicators into machine learning algorithms is impractical and empirical results have confirmed a severe drop in the performance of trading systems in such cases. New research into better ways of representing and abstracting data particularly textual data is needed. So far, work has been done in isolation e.g. Directional Changes (DC) adopts an event-based view on the data which is reminiscent of the work in Complex Event Processing (CEP) in which research aims at integrating CEP with ML techniques [3]. New types of information that can realistically represent the richness and diversity of existing institutional environments, regulations and trading behaviours need to be defined, including behavioural data e.g. one that reflects traders' emotions. This raises many challenges such as ensuring data quality issues as well as organizing metadata when combining data from multiple sources.
- **Developing more sophisticated machine learning techniques:** existing models will need to be adapted to include constraints, incentives and biases of decision makers (fund managers, traders, etc.) and their impact on market valuations. In addition, new models for quantifying risk are urgently required. For these reasons, interest in the use

of deep learning methods is growing, for their ability to detect and exploit interactions in the data that are not captured by any existing financial economic theory. In [18], the authors state that, since deep learning has been successfully applied in speech recognition, visual object recognition, object detection and many other domains, they expect to see similar improvements in the domain of time series predictions. In addition, the use of machine learning methods as a part of a bigger system in which other models co-exist has many practical advantages as illustrated by the growing trend to use ensemble methods [50]. The focus of the research becomes more on how to engineer such complex systems and conduct evaluation studies than on designing a new or customizing an existing machine learning technique.

- **Automatic trading strategy elaboration and execution**: the use of machine learning techniques in the elaboration of a trading strategy and its various parameters such as order size and venue choice, taking into account a wide range of goals, time horizons and benchmarks is still in its infancy. Extensive additional empirical work is needed in areas such as price forecasting according to different trading rules and horizons, researching the arbitrage between effective holding periods, determining transaction costs and elaborating the rules for order execution. Trying to capture individual/specific effects for a large number of financial instruments rises the number of predictors and the level of noise which reduces the performance of the trading system. Deeper empirical studies involving the hyper-optimization of all parameters are necessary. As more sophisticated and autonomous algorithms that can learn from their experiences (both positive and negative) in the markets are being proposed, the issue of maintaining and fine-tuning them to current market conditions becomes critical. It is expected that optimization of trading processes will also be increasingly automated, guided by applying machine learning techniques to learn from experiences captured in vast amounts of execution data.

These areas present a number of opportunities for productive multi-disciplinary collaborations between academia and practice, with academic fields including computer science, finance and econometrics. These studies should be guided by more insights into industry practices in this area, as the majority of the available information is currently coming from white papers and trade publications thar are sometimes characterized by narrow perspectives and commercial bias. The next step in this research will be to conduct an in-depth survey related to the use of machine learning techniques involving industry practitioners and augment the results of this preliminary survey with the results. This would give a more accurate picture of the state-of-art as well as help refine the areas of future research work identified in this paper.

Acknowledgements. We wish to acknowledge Plato Consulting and in particular Mike Bellaro for sponsoring this project. We also would like to thank Professor Carole Comerton-Forde, Professor Peter Gomber, Dr Giuseppe Nuti and Dr Kingsley Jones for their help and advice related to this project.

References

1. Azzini, A., Tettamanzi, A.: Automated trading on financial instruments with evolved neural networks. In: Proceedings of GECCO 2007: Genetic and Evolutionary Computation Conference, p. 2252 (2007). https://doi.org/10.1145/1276958.1277387
2. Barclays: Algo Wheels: 6 essential building blocks (2018). https://www.investmentbank. barclays.com/our-insights/algo-wheels-6-building-blocks.html#:~:text=An%20algo%20w heel%20is%20an,to%20mostly%20trader%2Ddirected%20flow. Accessed 24 Oct 2018
3. Bifet, A., Gavaldà, R., Holmes, G., Pfahringer, B.: Machine Learning for Data Streams, with Practical Examples in MOA. MIT Press (2018). https://doi.org/10.7551/mitpress/10654.001. 0001
4. BoE: Machine learning in UK financial services (2019). http://www.bankofengland.co.uk/ report/2019/machine-learning-in-uk-financial-services
5. Boonpeng, S., Jeatrakul, P.: Decision support system for investing in stock market by using OAA-neural network. In: 2016 Eighth International Conference on Advanced Computational Intelligence (ICACI), pp. 1–6 (2016). https://doi.org/10.1109/icaci.2016.7449794
6. Booth, A,. Gerding, E., Mcgroarty, F.: Automated trading with performance weighted random forests and seasonality. Expert. Syst. Appl. **41**, 3651–3661 (2014). https://doi.org/10.1016/j. eswa.2013.12.009
7. Cliff, D., Treleaven, P.: Technology trends in the financial markets: a 2020 vision, UK Government Office for Science's Foresight Driver Review on The Future of Computer Trading in Financial Markets – DR, 3 October 2010 (2010)
8. Cont, R., Kukanov, A.: Optimal order placement in limit order markets. Quant. Finance **17**(1), 21–39 (2017). https://doi.org/10.1080/14697688.2016.1190030
9. Dabous, F.T., Rabhi, F.A.: Information systems and IT architectures for securities trading. In: Seese, D., Weinhardt, C., Schlottmann, F. (eds.) Handbook on Information Technology in Finance. International Handbooks Information System, pp. 29–50. Springer, Heidelberg (2008). https://doi.org/10.1007/978-3-540-49487-4_2. ISBN 978-3-540-49486-7
10. Dai, Y., Zhang, Y.: Machine Learning in Stock Price Trend Forecasting, Stanford University (2013)
11. Das, S.: Text and context: language analytics in finance. Found. Trends® Finance **8**, 145–261 (2014). https://doi.org/10.1561/0500000045
12. Das, S.R.: The future of fintech. Financ. Manag. **48**(4), 981–1007 (2019). https://doi.org/10. 1111/fima.12297
13. Dash, R., Dash, P.: A hybrid stock trading framework integrating technical analysis with machine learning techniques. J. Finance Data Sci. **2** (2016). https://doi.org/10.1016/j.jfds. 2016.03.002
14. Day, M., Lee, C.: Deep learning for financial sentiment analysis on finance news providers. In: Kumar, R., Caverlee, J., Tong, H. (eds.) Proceedings of the 2016 IEEE/ACM International Conference on Advances in Social Networks Analysis and Mining, ASONAM 2016, pp. 1127–1134. IEEE (2016)
15. Joneston Dhas, J.L., Maria Celestin Vigila, S., Ezhil Star, C.: Forecasting of stock market by combining machine learning and big data analytics. In: Zelinka, I., Senkerik, R., Panda, G., Lekshmi Kanthan, P.S. (eds.) ICSCS 2018. CCIS, vol. 837, pp. 385–395. Springer, Singapore (2018). https://doi.org/10.1007/978-981-13-1936-5_41
16. Emir, Ş., Dincer, H., Timor, M.: A stock selection model based on fundamental and technical analysis variables by using artificial neural networks and support vector machines. Int. Rev. Econ. Finance **02**, 106–122 (2012)
17. Fan, J., Han, F., Liu, H.: Challenges of big data analysis. Natl. Sci. Rev. **1**(2), 293–314 (2014). https://doi.org/10.1093/nsr/nwt032

18. Fischer, T., Krauss, C.: Deep learning with long short-term memory networks for financial market predictions. Eur. J. Oper. Res. **270** (2017). https://doi.org/10.1016/j.ejor.2017.11.054
19. GlobeNewswire: InfoReach Includes Algo Wheel with TMS (2019). https://www.globenews wire.com/news-release/2019/12/17/1961861/0/en/InfoReach-Includes-Algo-Wheel-with-TMS.html. Accessed 19 Dec 2019
20. Goldkamp, J., Dehghanimohammadabadi, M.: Evolutionary multi-objective optimization for multivariate pairs trading. Expert. Syst. Appl. **135** (2019). https://doi.org/10.1016/j.eswa.2019.05.046
21. Gomber, P., Zimmermann, K.: Algorithmic trading in practice. In: The Oxford Handbook of Computational Economics and Finance, February 2018. https://doi.org/10.1093/oxfordhb/9780199844371.013.12
22. Gregoriou, G. (ed.): The Handbook of Trading: Strategies for Navigating and Profiting from Currency, Bond, and Stock Markets. McGraw-Hill, New York (2010)
23. Hu, Y., et al.: Application of evolutionary computation for rule discovery in stock algorithmic trading: a literature review. Appl. Soft Comput. **36**, 534–551 (2015). https://doi.org/10.1016/j.asoc.2015.07.008
24. Huang, B., Huan, Y., Xu, L.D., Zheng, L., Zou, Z.: Automated trading systems statistical and machine learning methods and hardware implementation: a survey. Enterp. Inf. Syst. **13**(1), 132–144 (2019). https://doi.org/10.1080/17517575.2018.1493145
25. Huck, N.: Large data sets and machine learning: applications to statistical arbitrage. Eur. J. Oper. Res. **278**(1), 330–342 (2019). https://doi.org/10.1016/j.ejor.2019.04.013
26. Investopedia: Understanding Order Execution (2020). https://www.investopedia.com/art icles/01/022801.asp. Accessed 24 Mar 2020
27. Jang, H., Lee, J.: Generative Bayesian neural network model for risk-neutral pricing of American index options. Quant. Finance, 1–17 (2018). https://doi.org/10.1080/14697688.2018.1490807
28. Garcia-Lopez, F., Batyrshin, I., Gelbukh, A.: Analysis of relationships between tweets and stock market trends. J. Intell. Fuzzy Syst. **34**, 1–11 (2018). https://doi.org/10.3233/JIFS-169515
29. JPMorgan: JPMorgan e-Trading 2020 survey (2020). https://www.jpmorgan.com/global/mar kets/e-trading-2020
30. Kampouridis, M., Barril Otero, F.: Evolving trading strategies using directional changes. Expert Syst. Appl. **73**, 145–160 (2017). https://doi.org/10.1016/j.eswa.2016.12.032
31. Li, X., Wang, C., Dong, J., Wang, F., Deng, X., Zhu, S.: Improving stock market prediction by integrating both market news and stock prices. In: Hameurlain, A., Liddle, S.W., Schewe, K.-D., Zhou, X. (eds.) DEXA 2011. LNCS, vol. 6861, pp. 279–293. Springer, Heidelberg (2011). https://doi.org/10.1007/978-3-642-23091-2_24
32. Li, X., et al.: Empirical analysis: stock market prediction via extreme learning machine. Neural Comput. Appl. **27**(1), 67–78 (2014). https://doi.org/10.1007/s00521-014-1550-z
33. Lv, D., Huang, Z., Li, M., Xiang, Y.: Selection of the optimal trading model for stock investment in different industries. PLoS ONE **14**(2) (2019). https://doi.org/10.1371/journal.pone.0212137
34. Narang, R.: Inside the Black Box: The Simple Truth About Quantitative Trading. Wiley Finance (2009)
35. Paiva, F., Cardoso, R., Hanaoka, G., Duarte, W.: Decision-making for financial trading: a fusion approach of machine learning and portfolio selection. Expert Syst. Appl. **115**, 635–655 (2019)
36. Pimprikar, R., Ramachandran, S., Senthilkumar, K.: Use of machine learning algorithms and Twitter sentiment analysis for stock market prediction. Int. J. Pure Appl. Math. **115**(6), 521–526 (2017)

37. Pole, A.: Statistical Arbitrage: Algorithmic Trading Insights and Techniques. Wiley Finance (2007)
38. Porshnev, A., Redkin, I., Shevchenko, A.: Machine learning in prediction of stock market indicators based on historical data and data from twitter sentiment analysis. In: 2013 IEEE 13th International Conference on Data Mining Workshops 2013, pp. 440–444 (2013). https://doi.org/10.1109/icdmw.2013.111
39. Preda, A.: Noise: Living and Trading in Electronic Finance, 1st edn. University of Chicago Press, Chicago (2017)
40. Psomadelis, W.: Solving Execution; Contextual Analysis, Intelligent Routing and the Role of the Algo Wheel, Global Trading, 19 May 2019. https://www.fixglobal.com/home/solving-execution-contextual-analysis-intelligent-routing-and-the-role-of-the-algo-wheel/
41. Reserve Bank of Australia: Electronic Trading in Australian Financial Markets, RBA Bulletin (2001). https://www.rba.gov.au/publications/bulletin/2001/dec/2.html. Accessed Dec 2001
42. Robertson, C.S., Rabhi, F.A., Peat, M.: A service-oriented approach towards real time financial news analysis. In: Lin, A., Foster, J., Scifleet, P. (eds.) Consumer Information Systems: Design, Implementation and Use. IGI Global, 2012 (2012)
43. Rodríguez-González, A., et al.: Improving trading systems using the RSI financial indicator and neural networks. In: Kang, B.-H., Richards, D. (eds.) PKAW 2010. LNCS (LNAI), vol. 6232, pp. 27–37. Springer, Heidelberg (2010). https://doi.org/10.1007/978-3-642-15037-1_3
44. Rundo, F., Trenta, F., Battiato, S., Stallo, A.: Advanced markov-based machine learning framework for making adaptive trading system (2019). https://doi.org/10.3390/computation7010004
45. Schulenburg, S., Ross, P.: Explorations in LCS models of stock trading. In: Lanzi, P.L., Stolzmann, W., Wilson, S.W. (eds.) IWLCS 2001. LNCS (LNAI), vol. 2321, pp. 151–180. Springer, Heidelberg (2002). https://doi.org/10.1007/3-540-48104-4_10
46. Schumaker, R.P., Chen, H.: A discrete stock price prediction engine based on financial news. Computer 43(1), 51–56 (2010). https://doi.org/10.1109/mc.2010.2
47. Thursfield, J.: Algo wheel adoption picks up in derivs – expert. Global Investor (2019)
48. Vella, V., Lon Ng, W.: A dynamic fuzzy money management approach for controlling the intraday risk-adjusted performance of AI trading algorithms. Intell. Syst. Account. Finance Manag. 22(2), 153–178 (2015)
49. Weber, J., Riera, M.: Welcome to the new world of equity trade execution: MiFID II, algo wheels and AI. Targeted News Service, Greenwich Associates News Release, 23 April 2019 (2019)
50. Weng, B., Lu, L., Wang, X. Megahed, F., Martinez, W.: Predicting short-term stock prices using ensemble methods and online data sources. Expert. Syst. Appl. 112 (2018). https://doi.org/10.1016/j.eswa.2018.06.016
51. Yang, S.Y., Qiao, Q., Beling, P.A., Scherer, W.T., Kirilenko, A.A.: Gaussian process-based algorithmic trading strategy identification. Quant. Finance 15(10), 1683–1703 (2015). https://doi.org/10.1080/14697688.2015.1011684

Using Machine Learning to Predict Short-Term Movements of the Bitcoin Market

Patrick Jaquart[✉], David Dann, and Christof Weinhardt

Karlsruhe Institute for Technology, Kaiserstraße 89-93, 76133 Karlsruhe, Germany
{patrick.jaquart,david.dann,christof.weinhardt}@kit.edu

Abstract. We analyze the predictability of the bitcoin market across prediction horizons ranging from 1 to 60 min. In doing so, we test various machine learning models and find that, while all models outperform a random classifier, recurrent neural networks and gradient boosting classifiers are especially well-suited for the examined prediction tasks. We use a comprehensive feature set, including technical, blockchain-based, sentiment-/interest-based, and asset-based features. Our results show that technical features remain most relevant for most methods, followed by selected blockchain-based and interest-based features. Additionally, we find that predictability increases for longer prediction horizons. Although a quantile-based long-short trading strategy generates monthly returns of up to 31% before transaction costs, it leads to negative returns after taking transaction costs into account due to the particularly short holding periods.

Keywords: Bitcoin · Machine learning · Financial market prediction · Neural networks · Gradient boosting · Random forests

1 Introduction

Bitcoin is a digital currency, introduced in 2008 by Nakamoto [1]. In this study, we analyze the short-term predictability of the bitcoin market. Therefore, we utilize a variety of machine learning methods and consider a comprehensive set of potential market-predictive features.

Empirical asset pricing is a major branch of financial research. In contrast to the well-established research stream of equity pricing, which has yielded a substantial number of potentially market-predictive factors [2], research on the pricing of cryptocurrencies is not yet mature. In particular, the short-term predictability of the bitcoin market has not been analyzed comprehensively. Furthermore, most researchers have only considered technical features and have not analyzed the feature importance of the employed machine learning models [3]. We tackle this research gap by comparatively analyzing different machine learning models for predicting market movements of the most relevant cryptocurrency—bitcoin. With a market capitalization of around 170 billion US dollar (July 2020),

© Springer Nature Switzerland AG 2020
B. Clapham and J.-A. Koch (Eds.): FinanceCom 2020, LNBIP 401, pp. 21–40, 2020.
https://doi.org/10.1007/978-3-030-64466-6_2

bitcoin represents about 63% of the cryptocurrency market [4]. In this context, our overarching research question is:

RQ: Is it possible to predict short-term movements of the bitcoin market?

Our study provides two main contributions. First, we contribute to the literature by systematically comparing the predictive capability of different prediction models (e.g, recurrent neural networks, gradient boosting classifiers), feature sets (e.g., technical, blockchain-based), and prediction horizons (1–60 min). Thereby, our study establishes a thorough benchmark for the predictive accuracy of short-term bitcoin market prediction models. Second, despite the models' ability to create viable bitcoin market predictions, our results are in line with the efficient market hypothesis, as the returns generated by the employed trading strategy (based on the short-term predictions) are not able to compensate for associated transaction costs.

2   Related Work

Financial market prediction is an extensively-studied branch of financial research [3]. There is mixed evidence regarding the predictability and efficiency of financial markets [5,6]. An established approach to analyze return-predictive signals is to conduct regressions on possible signals to explain asset returns [7]. However, linear regressions cannot incorporate a large number of features flexibly and impose strong assumptions on the functional form of how signals affect the market. Therefore, machine learning methods, which often do not impose those restrictions, have been increasingly applied for financial market prediction [8,9]. Among those, neural network-based methods are expected to be particularly well-suited, as they were already described to be the dominant method for predicting the dynamics of financial markets [10].

2.1   Market Efficiency and Financial Market Prediction

Theory on Market Efficiency. Within efficient financial markets, prices reflect all available information and are not predictable in order to earn excess returns. To determine the degree of efficiency of a market, Fama [5] defines a formal three-level framework—weak, semi-strong, and strong form market efficiency. In weak form efficient markets, prices reflect all information about past prices, whereby in semi-strong form efficient markets, prices reflect all publicly available information. In strong form efficient markets, prices additionally reflect all private information. While regulators aim to prevent investors from profiting from private information, it is generally agreed upon that major financial markets are semi-strong form efficient [11]. However, Grossman and Stiglitz [12] argue that market efficiency may not constitute a constant state of equilibrium over time. If information is costly and prices consistently reflect all available information, informed traders will stop to acquire information, which leads market prices to deviate from fundamental asset values. Additionally, there is evidence in the financial literature for a large number of potential market anomalies. Green et al. [13], for instance, identify more than 330 different empirically-found return-predictive signals for the US stock market that have been published between 1970 and 2010. Similarly, Lo [6] formulates the adaptive markets hypothesis, according to which markets may be temporarily inefficient. Thereby, the duration of the temporal inefficiency is influenced by the degree of competition within a market and limits to arbitrage, since informed traders exploit existing inefficiencies so that prices reflect all available information again. Against this backdrop, it remains an open question, whether return-predictive signals constitute market anomalies or represent reasonably priced risk factors. Also, some of the most prominent market anomalies have disappeared after publication [14], which indicates that part of the published return-predictive signals either have only existed in the sample period or have been erased due to traders adopting strategies for exploitation. Green et al. [13] infer that a unified model of market efficiency or inefficiency should account for persistent empirically identified return-predictive signals.

Bitcoin Market Efficiency. Several findings in the financial literature [15, 16] indicate that bitcoin could be part of a new asset class. Therefore, findings regarding weak form efficiency of other financial markets may not hold for the bitcoin market. Several researchers examine the degree of bitcoin market efficiency using different time horizons. First, Urquhart [17] investigates daily bitcoin prices (August 2010 to July 2016). He finds that the bitcoin market is not even weak form efficient. However, splitting the study period reveals that the bitcoin market becomes increasingly efficient over time. Revisiting this data, Nadarajah and Chu [18] find that a power transformation of the used bitcoin returns satisfies the weak form efficient market hypothesis. Similarly, Bariviera [19] examines daily bitcoin prices (August 2011 to February 2017) and shows that the bitcoin market is not weak form efficient before 2014, but becomes weak form efficient after 2014. Vidal-Tomás and Ibañez approach the question of semi-strong form bitcoin market efficiency from an event study perspective [20]. With data (September 2011 to December 2017) on news related to monetary policy changes and bitcoin, they show that the bitcoin market does not react to monetary policy changes but becomes increasingly efficient concerning bitcoin-related events. Testing for the adaptive markets hypothesis, Khuntia and Pattanayak [21] analyze daily bitcoin prices (July 2010 to December 2017), finding evidence for an evolving degree of weak form market efficiency. They conclude that this finding constitutes evidence that the adaptive market hypothesis holds for the bitcoin market.

Summarizing, there is mixed evidence among scholars regarding the efficiency of the bitcoin market. However, most researchers find that the bitcoin market has become more efficient over the years. An increasing degree of market efficiency seems intuitive, as the bitcoin market has proliferated since its inception and, therefore, has become increasingly competitive.

2.2   Bitcoin Market Prediction via Machine Learning

Jaquart et al. [3] extensively review the literature on bitcoin market prediction via machine learning. They analyze the body of literature with regards to applied machine learning methods, return-predictive features, prediction intervals, and prediction types. The reviewed literature utilizes both classification and regression models approximately equally often, while regression models are used slightly more frequently. Due to the use of different time horizons, targets and feature variables, parameter specifications, and evaluation metrics, the comparison of prediction models across different papers often remains not possible. On the other hand, comparisons within the same paper often avoid these shortcomings, and remain especially relevant. Based on the latter, Jaquart et al. [3] outline that especially recurrent neural networks yield promising results regarding bitcoin market predictions (e.g., [22,23]). Furthermore, they group the utilized return-predictive features into four major categories—technical, blockchain-based, sentiment-/interest-based, and asset-based features. Thereby, technical features describe features that are related to historical bitcoin market data (e.g., bitcoin returns). Blockchain-based features denote features related to the bitcoin blockchain (e.g., number of bitcoin transactions). Sentiment-/interest-based features describe features that are related to sentiment and internet search volume of bitcoin (e.g., bitcoin Twitter sentiment). Asset-based features are features that are related to financial markets other than the bitcoin market (e.g., gold returns, returns of the MSCI World index).

So far, only few researchers utilize features of *all* established feature categories. Besides, the particular feature importance across different models has received little academic attention yet. Next, the vast majority of researchers construct their models using daily prediction horizons [3]. Furthermore, only few scholars [24,25] contrast different prediction periods against each other. Consequently, the bitcoin market dynamics concerning prediction intervals of less than one hour are not fully understood yet.

3   Methodology

To tackle the previously-outlined research gap, we systematically evaluate different prediction models, features, and horizons. Therefore, we implement data gathering, preprocessing, and model-building using the Python programming language and the libraries TensorFlow, scikit-learn, and XGBoost.

3.1   Data

We use data from Bloomberg, Twitter, and Blockchain.com ranging from March 2019 to December 2019. Regarding Bloomberg, our data set includes minutely price data for bitcoin, gold, oil and minutely levels for the total return variants of the indices MSCI World, S&P 500, and VIX. All prices and index levels are denoted in US dollar (USD). Furthermore, the data set includes minutely exchanges rates relative to the US Dollar for the currencies euro (EUR/USD), Chinese yuan (CNY/USD), and Japanese yen (JPY/USD). From Blockchain.com, the data set includes minutely data for the number of bitcoin transactions and growth of the mempool (i.e., storage of not-yet validated bitcoin transactions). Last, the data set includes sentiment data of all English Twitter tweets in the given period that include the hashtag bitcoin ("#bitcoin").

3.2   Generation of Training, Validation and Test Sets

We convert all timestamps to Coordinated Universal Time (UTC) and create a data set for each prediction problem by aggregating features and target variable. Most bitcoin trading venues allow for continuous trading of bitcoin but for the minutely Bloomberg bitcoin price series, there is a gap in the time series on weekends, which we exclude from our analysis. Since the utilized asset-based features are related to assets that are mainly traded on weekdays, we consider this procedure to be reasonable. Due to the 7-day bitcoin return feature, we require for every observation seven days of history to calculate the complete feature set. Therefore, our final data sample spans a time range of 9 months, namely from March 11, 2019 to December 10, 2019. We use the first 5/9 of the data (approximately five months) to generate a training set. The subsequent 1/9 of the data (approximately one month) forms the validation set for hyperparameter tuning, including regularization techniques, such as early stopping [26]. The remaining 1/3 of data (approximately 3 months) is used to test our models and obtain a representative out-of-sample prediction accuracy.

3.3   Features

We employ features from all four major feature categories identified by Jaquart and colleagues [3], as listed in Table 1. For all prediction models, we calculate minutely-updated feature values. Depending on whether the prediction model has a memory state, we further aggregate these feature values.

For the *technical* and *asset-based* features, returns are given by

$$r^a_{t,t-k} = \frac{p^a_t}{p^a_{t-k}} - 1, \tag{1}$$

where p^a_t is defined as the price of asset a at time t and k represents the number of periods over which the return is calculated. We obtain minutely-updated values for the selected *blockchain-based* features. *Sentiment-/interest-based* features are generated from the collected tweets. We only keep tweets that do not contain pictures or URLs, since the use of textual sentiment analysis is not able to capture all information contained in multimodal tweets [27]. Following the suggestions of Symeonidis et al. [28], we apply various pre-processing techniques to the collected tweet texts. First, we remove usernames, non-English characters, and additional whitespace. Second, we replace contractions (e.g., replace "isn't" with "is not"). Last, we apply lemmatization to the tweet texts to replace inflected word forms with respective word lemmas (e.g., replace "bought" with "buy"). Next, we make use of the Google Natural Language API to generate sentiment and estimates of strength of emotion for each tweet. For every minutely instance, we calculate three different features: First, the number of bitcoin tweets published in last minute as a proxy for the overall interest in bitcoin. Second, the sum of sentiment scores of all tweets published in the previous minute. Third, the weigh the sentiment scores by the strength of emotion per tweet. Feature two and three depend on the sentiment expressed towards bitcoin but differ in the applied weighting scheme. While traditional prediction methods fail more often when predictors are highly correlated, machine learning models appear well-suited for these cases utilizing various variable selection methods [9].

Table 1. Overview of the utilized features.

Technical	
Bitcoin returns	
Asset-based	
MSCI World returns	Crude Oil WTI returns
SP 500 returns	EUR/USD returns
VIX returns	CNY/USD returns
Gold returns	JPY/USD returns
Blockchain-based	
Number of bitcoin transactions	Mempool growth
Sentiment-/Interest-based	
Twitter sentiment	Number of tweets
Twitter sentiment weighted with strength of emotion	

Feature Set for Models with Memory Function. For the machine learning models with a memory function (i.e., LSTM and GRU), we create time series for all features listed in Table 1. To facilitate model training, all feature values are standardized based on the properties of the specific feature in the training set [26].

Following feature standardization, we create time series from the 120 most recent, minutely features values. For the employed technical and asset-based features, the time series consist of the latest one-minute returns. However, for bitcoin, we create an additional time series by also calculating the one-week bitcoin returns (Eq. 1) for each of the most recent 120 min to give the models information about the longer-term status of the bitcoin market. Conclusively, the input for the memory models consists of 15 different time series, whereby each of the time series consists of 120 minutely time steps.

Feature Set for Models without Memory Function. The prediction models without memory function (i.e., feedforward networks, random forests, gradient boosting classifiers, and logistic regressions), require input form of a one-dimensional vector with one observation per feature. Therefore, we create additional features by aggregating the 120-min history of the feature classes to also give the employed no-memory models temporal information about the feature values. In line with Takeuchi and Lee [29] and Krauss et al. [8], we choose a more granular resolution for the most recent feature history. Specifically, we choose the following set of intervals, j, to aggregate the feature history: $j \in \{(0, 1],$ $(1, 2], (2, 3], (3, 4], (4, 5], (5, 10], (10, 20], (20, 40], (40, 60], (60, 80], (80, 100],$ $(100, 120]\}$, whereby these intervals describe the minutes before a prediction is made. For the aggregation process, blockchain-based features, as well as sentiment-/interest-based features are summed up across the respective intervals. For the aggregated technical and asset-based features, we calculate multiperiod returns over the respective intervals (Eq. 1). We build these intervals for all features used for the feature sequences of the memory models, except for the one-week bitcoin return, since this time series naturally exhibits low variation over 120 consecutive minutes. As for the memory models, we standardize all features based on the feature properties within the training set. Consequently, our feature set for the prediction models without memory function consists of $14 \times 12 + 1 = 169$ different features.

3.4 Targets

We formulate a binary classification problem for four different prediction horizons. For every observation, the target class c_m is formed based on the return over the next m minutes, with $m \in \{1, 5, 15, 60\}$.

We place observations, for which the return over the next m minutes is greater than or equal to the median m-minute return of all training set observations, in class one and all other observations in class zero. With regard to Eq. 1, the target class y_t^m, is given by,

$$y_t^m = \begin{cases} 1, & \text{if } r_{t+m,t}^{bitcoin} \geq Median(r_{u+m,u}^{bitcoin}) \forall u \in \{train\} \\ 0, & \text{otherwise} \end{cases}, \qquad (2)$$

where *train* denotes all time timestamps in the training set.

Creating classes directly from the training set ensures that the prediction models are trained on equally balanced proportions and are not subject to a bias towards one specific class. During prediction, a model returns the probability for an observation to belong to a specific class.

3.5 Prediction Models

With our set of models for bitcoin market prediction, we benchmark neural networks with and without memory components, tree-based models, regression models, and ensemble models against each other. Apart from the ensemble models, all these models have already been applied to the domain of bitcoin market predictions [3]. Besides the following described models, we use a logistic regression model (LR) as a benchmark.

Neural Networks. The structure and intended behavior of artificial neural networks is inspired by the functionality of the human brain. In analogy to the structure of a brain, which consists of billions of highly interconnected neurons, artificial neural networks consist of various, highly connected nodes. All networks employed contain individual dropout layers and are trained with the established Adam optimizer [30] to minimize the binary cross-entropy loss, and use the early-stopping method to improve the level of generalization [26].

Feedforward Neural Network. Feedforward neural networks (FNN) are a basic type of neural networks [31]. FNNs represent a directed acyclic graph in which processed information flows exclusively in one direction. FNNs consist of three types of layers: One input layer, capturing the input information, a variable number of hidden layers, and one output layer, determining the network's final classification. The final classification is dependent on the activation of nodes in preceding layers. The activation of each node in all layers is determined by a previously-assigned (commonly non-linear) activation function. Figure 1 (bottom) describes the architecture of the applied FNNs.

LSTM and GRU. Long short-term memory (LSTM) and gated recurrent unit (GRU) networks belong to the category of gated recurrent neural networks (RNNs). RNNs drop FFN's condition of acyclic graphs. Hochreiter and Schmidhuber introduce the LSTM architecture in the late 1990's [32] with a specific focus on long-term memorization of information in sequential data. Their architecture replaces the nodes in the hidden layers with memory blocks. Each block usually consists of one memory cell and varying number of gates, which can manipulate the internal values of the cell. The original LSTM has two different gates: an input and an output gate. Each gate utilizes a sigmoid activation function. GRUs differ from the LSTMs insofar, as they use one unified gate unit to control the forget and the update gate simultaneously. Although the number of learnable parameters of GRUs is thereby smaller than that of LSTMs, their performance in various domains is comparable [33]. Figure 1 (top) outlines the architecture of the applied RNNs.

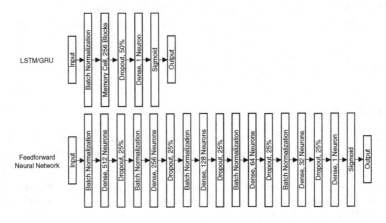

Fig. 1. Architecture of applied feedforward neural networks (bottom) and recurrent neural networks (top). Note: For the recurrent networks, the memory cell is either an LSTM cell or a GRU cell.

Tree-Based Models. Tree-based models use a decision tree to learn attribute-class relationships, which are fast to train and well interpretable. However, they are prone to overfit.

Random Forest. Introduced by Ho in 1995 [34], random forests (RF) aim to overcome tree-based models' tendency to overfit by means of an ensemble method. Here, multiple decision trees are generated, and each of them is trained on different parts of the training data. The output of the final model for overall classification is the average output of all individual trees. The random forest applied is subject to the parameterization of 100 trees and a minimum number of instances per leaf of 20%. For all remaining parameters, we use the default values of the Python scikit-learn library.

Gradient Boosting Classifier. Similar to random forests, gradient boosting classifiers (GBC) leverage the input of multiple decision trees [35]. In addition, boosted classifiers also train individual weights for all included models. In this way, the output classification of the better-adapted models is weighted more strongly in the model's final classification decision. In our analysis, we use the extreme gradient boosted (XGBoost) trees, parameterized with a binary logistic objective function to build a gradient boosting classifier, whereby individual trees have a max-depth of 1.

Ensemble Models. Similar to random forest and gradient boosting, ensemble models rely on the classification output of multiple models. However, in an ensemble, different model-types (e.g., neural networks and tree-based models) can be combined into a meta-model. The output of the meta-model constitutes the averaged predictive probability vector of all models included. In this way, method-specific misclassifications should be "overruled" by the other models in the ensemble. We apply an meta-model consisting of all individual models.

3.6   Evaluation

The prediction models are evaluated and analyzed regarding various aspects. First, we compare the models on a prediction level. Second, we analyze and compare feature importance for each model and prediction target. Third, we examine economic implications of our bitcoin market predictions.

Forecast Evaluation. We compare the forecasts of our prediction models based on the predictive accuracy on the test set. Furthermore, we compare the significance of the differences in model-prediction based on the Diebold-Mariano test [36]. Additionally, we estimate the probability that the models make predictions by chance. If the true accuracy of a binary classification model is 50%, the number of correctly classified targets follows the distribution

$$X \sim B(n = \#test, p = 0.5, q = 0.5), \tag{3}$$

where $\#test$ is the number of observations in the test sample (e.g. 94834 for the 1-min horizon). Based on this binomial distribution, we calculate the probabilities that a prediction model has a true probability of 50%.

Feature Importance. The feature importance for all models is determined by the measure of permutation feature importance [37]. This ensures comparability between the resulting importance scores across all models. We randomly permute every feature vector with a random standard normally distributed vector and calculate the decrease in prediction accuracy, which we interpret as feature importance. A high decrease in prediction accuracy implies that the model strongly relies on this feature for its predictions. To decrease the impact of randomness on the results, we average the permutation feature importance across a set of 10 different random seeds, $s \in \{0, 1, 2, 3, 4, 5, 6, 7, 8, 9\}$. In the rare case that the random permutation increases the prediction accuracy, we set this feature's importance to zero. To enhance interpretability, we follow Gu et al.'s recommendations [9] and normalize the feature importances in a way that all feature importance scores for a prediction model sum to one.

Trading Strategy. We analyze the economic implications of the bitcoin market predictions by testing a straightforward trading strategy. To approximate an ex ante trading strategy, we calculate the 99%-quantiles of all the classes' probabilities from the training set predictions. If, for instance, the predicted probability in the test set for class 1 is higher than the respective threshold probability, we take a long position. Vice versa, we take a short position, if a predicted class probability for class 0 is above than the respective threshold probability. At the end of the prediction interval, the position is closed. We calculate the return of this strategy before and after transaction costs. Similar to Fischer and colleagues [38], we assume round-trip transaction costs of 30 basis points (bps).

4   Results

Predictive Accuracy. We compare the model predictions based on the accuracy score. First, as shown in Table 2, we find that all tested models' predictive accuracy is beyond the 50% threshold. Furthermore all models have a probability of less than $4.18E-12$ for a true accuracy of 50% (see Table 4). Second, we find that the average prediction accuracy monotonically increases for longer prediction periods. Third, we find that RNNs or GBCs constitute the best-performing methods across all prediction periods. Specifically, the GRU performs best on the 1-min prediction horizon. Diebold-Mariano tests (Table 5) reveal that GRU predictions are more accurate than the predictions of the LR, GBC, and the ensemble model ($\alpha : 5\%$). On the 5-min and 15-min horizon, the GBC model has the highest predictive accuracy, yielding more accurate forecasts than all other models, apart from the meta-ensemble on the 5-min horizon. The LSTM model makes the most accurate predictions on the 60-min horizon. These predictions are statistically more accurate than the predictions of the other models. However, the difference is not significant compared to GBC, RF, and the meta-ensemble.

Table 2. Prediction accuracy of the machine learning models for the different prediction horizons.

Model	Accuracy			
	1-min predictions	5-min predictions	15-min predictions	60-min predictions
GRU	0.515553	0.523180	0.527414	0.537989
LSTM	0.513887	0.521840	0.528079	0.558557
FFN	0.514415	0.518580	0.518728	0.533011
LR	0.511272	0.517926	0.519595	0.538552
GBC	0.511093	0.529268	0.537282	0.557026
RF	0.514183	0.523085	0.532211	0.556313
E (All)	0.513740	0.527622	0.532855	0.557334

Feature Importance. On the 1-min prediction horizon, the predominant feature for both RNNs is the minutely bitcoin return time series with a relative importance of about 90%. For the LSTM, the minutely bitcoin return also constitutes the most important feature (relative feature importance of about 50%) on the 60-min prediction horizon. For the GRU, however, the number of bitcoin transactions is the most important feature on this time horizon. It becomes apparent that, for longer time horizons, additional time series besides the minutely bitcoin returns are increasingly relevant for both RNNs. Among those are the number of transactions per second, the weekly bitcoin returns, and the weighted sentiment scores.

Next, subsequent analysis of our models without memory function provides further insights into the temporal feature importance distribution. The most important feature for the RF and the GBC on the 1-min prediction horizon is the bitcoin return over the last minute. On the 60-min prediction horizon, the bitcoin returns of the period from 40 min to 20 min constitute the most important feature for both tree-based models.

Similar to the finding for the RNNs, the relative importance of the predominant feature drops for the GBC for longer prediction horizons (60-min horizon: 70%, 1-min horizon: 30%). Besides technical features, mainly blockchain-based features (e.g., transactions per second, mempool size growth), as well as sentiment-/interest-based features (e.g. number of tweets) remain important for the tree-based models. Last, for the FFN and LR, feature importance is distributed along several features, which may be explained by the rather shallow parameterization of the tree-based models. We provide graphical representations of all feature importance in Appendix B.

Trading Strategy. Table 3 lists the results of our quantile-based trading strategy before transaction costs. Since the threshold class probabilities are calculated based on predictions for the training set, the number of trades varies between methods and prediction intervals. The results of the trading strategy yield three key takeaways. First, there is a rather large variance in trading results between the different prediction models. Higher predictive model accuracy does not necessarily translate into better trading results. We explain this by the fact that we do not set up our prediction problem to *optimize* trading performance, but rather to *predict* directional market movements. Additionally, based on our trading strategy, only a small proportion of observations is traded, which presumably increases variance. Second, the average return per trade tends to increase with longer prediction intervals. Third, considering transaction costs of 30 bps per round-trip, trading performance becomes negative for all methods. These negative returns may be explained by the models' short-term prediction horizons. Based on the transaction costs, making 1000 trades would cause transaction costs of 300%.

Table 3. Trading returns (*TR*) and number of trades (*#trades*) for the long-short 1%-quantile strategy before transaction costs over the three months of testing data.

Model	1-min predictions		5-min predictions		15-min predictions		60-min predictions	
	TR	#trades	TR	#trades	TR	#trades	TR	#trades
GRU	−0.0319	535	−0.1337	531	−0.1039	2730	0.836	1815
LSTM	−0.0147	575	−0.0117	413	0.4134	1783	0.9412	1326
FFN	−0.0061	1322	0.0570	756	0.2533	1247	0.1727	592
LR	0.0599	664	0.3405	822	0.2059	881	−0.8732	929
GBC	0.0637	1305	0.0097	1601	0.4749	1971	0.3155	1602
RF	0.0147	561	0.1341	608	0.3382	443	0.6265	654
E (All)	0.0265	662	0.0512	708	0.2629	1385	0.3517	1034

5   Discussion

This study shows that it is possible to make viable short-term market predictions for the most prominent cryptocurrency, bitcoin. However, the forecasting accuracy of slightly over 50% indicates that the bitcoin market predictability is somewhat limited. This may be due to multiple reasons, for instance, an immediate market reaction to the utilized features or a potentially large amount of information beyond these features that influence the bitcoin market. Furthermore, our results are consistent with the findings that the bitcoin market has become more efficient over the last years [17,20,21]. A limited bitcoin market predictability is comparable to findings related to the market predictability of other financial assets, such as stocks [8,9]. Since trading results based on the market predictions are negative after transaction costs, our work does not represent a challenge to bitcoin market efficiency. However, in this study, we aim to analyze the predictability of the bitcoin market movement and do not train our models to maximize trading performance. Nevertheless, our results indicate that trading strategies based on market predictions should preferably be implemented based on models with more extended prediction horizons. This would correspond to longer holding periods, for which the relative impact of transaction costs is presumably lower. Complementary, our finding that predictive accuracy increases for longer prediction horizons paves the path for further research opportunities. Next, we find that RNN and GBC models are especially well-suited for predicting the short-term bitcoin market, which may be explained by the ability of these models to clearly distinguish in the weighting of features, mainly relying on a set of few features. Technical features appear to be most influential, followed by blockchain-based and sentiment-/interest-based features. However, the exact source of predictive power for these features remains unclear. Possible sources of explanations may be theoretical equilibrium models. For instance, Biais et al. [39] determine a quasi-fundamental value for bitcoin based on, for instance, transactional costs and benefits. Some of the used features (e.g., transactions per second) may partially approximate these factors. Furthermore, it could be analyzed whether market anomalies, such as the momentum effect [40], exist within the bitcoin market. Furthermore, future researchers may examine whether behavioral financial market biases (e.g., the disposition effect [41]) are more pronounced for bitcoin, as it does not exhibit a fundamental value in the traditional sense.

6 Conclusion

In our empirical analysis, we analyze the short-term predictability of the bitcoin market, leveraging different machine learning models on four different prediction horizons. We find that all tested models make statistically viable predictions. We can predict the binary market movement with an accuracy of 51.1% to 55.7%, whereby the predictive accuracy tends to increase for longer forecast horizons. We identify that especially recurrent neural networks, as well as tree-based gradient boosting classifiers, appear to be well-suited for this prediction task. Comparing feature groups of technical, blockchain-based, sentiment-/interest-based, and asset-based features shows that for most methods, technical features remain prevalent. For a longer prediction horizon, the relative importance appears to spread across multiple features, including transactions per second and weighted sentiment. Feature analysis for the models without memory function reveals that for longer prediction horizons, less recent technical features become more relevant. We test a quantile-based trading strategy, which yields up to 94% return over three months before transaction costs. However, these returns cannot compensate for the transaction costs due to the low holding periods and correspondingly frequent trading activities.

Acknowledgements. The authors gratefully acknowledge financial support from the ForDigital Research Alliance.

Appendix

A   Supplemental Tables

Table 4. Probability of a true model accuracy of 50% derived from the binomial distribution described in Eq. 3.

Method	Probabilities			
	1-min predictions	5-min predictions	15-min predictions	60-min predictions
GRU	4.88E−22	1.62E−46	3.93E−64	2.34E−120
LSTM	5.98E−18	1.59E−41	3.53E−67	9.98E−283
FFN	3.40E−19	1.32E−30	5.04E−31	1.96E−91
LR	1.92E−12	1.26E−28	8.99E−34	6.90E−124
GBC	4.18E−12	6.67E−73	9.47E−117	2.93E−268
RF	1.22E−18	3.75E−46	1.03E−87	1.17E−261
E (All)	1.31E−17	3.61E−65	3.60E−91	3.84E−271

Table 5. Diebold-Mariano test p-values to reject the null hypothesis towards the alternative hypothesis that the forecast of model i on the test sample is more accurate than the forecast of model j.

1-min predictions							
j							
i	GRU	LSTM	FFN	LR	GBC	RF	E (All)
GRU	–	0.1107	0.2598	0.0094	0.0006	0.1613	0.0476
LSTM	0.8893	–	0.6126	0.0711	0.0323	0.5747	0.4522
FFN	0.7402	0.3874	–	0.0322	0.0398	0.4438	0.3442
LR	0.9906	0.9289	0.9678	–	0.4610	0.9527	0.9354
GBC	0.9994	0.9677	0.9602	0.5390	–	0.9958	0.9917
RF	0.8387	0.4253	0.5562	0.0473	0.0042	–	0.3506
E (All)	0.9524	0.5478	0.6558	0.0646	0.0083	0.6494	–

5-min predictions							
j							
i	GRU	LSTM	FFN	LR	GBC	RF	E (All)
GRU	–	0.2161	0.0064	0.0015	0.9997	0.4794	0.9985
LSTM	0.7839	–	0.0242	0.0111	1.0000	0.7753	1.0000
FFN	0.9936	0.9758	–	0.3442	1.0000	0.9971	1.0000
LR	0.9985	0.9889	0.6558	–	1.0000	0.9976	1.0000
GBC	0.0003	0.0000	0.0000	0.0000	–	0.0000	0.1251
RF	0.5206	0.2247	0.0029	0.0024	1.0000	–	0.9988
E (All)	0.0015	0.0000	0.0000	0.0000	0.8749	0.0012	–

(*continued*)

Table 5. (*continued*)

15-min predictions

j							
i	GRU	LSTM	FFN	LR	GBC	RF	E (All)
GRU	–	0.6728	0.0000	0.0000	1.0000	0.9972	0.9999
LSTM	0.3272	–	0.0000	0.0000	1.0000	0.9967	0.9999
FFN	1.0000	1.0000	–	0.7072	1.0000	1.0000	1.0000
LR	1.0000	1.0000	0.2928	–	1.0000	1.0000	1.0000
GBC	0.0000	0.0000	0.0000	0.0000	–	0.0013	0.0021
RF	0.0028	0.0033	0.0000	0.0000	0.9987	–	0.6664
E (All)	0.0001	0.0001	0.0000	0.0000	0.9979	0.3336	–

60-min predictions

j							
i	GRU	LSTM	FFN	LR	GBC	RF	E (All)
GRU	–	1.0000	0.0047	0.6216	1.0000	1.0000	1.0000
LSTM	0.0000	–	0.0000	0.0000	0.2018	0.0614	0.1698
FFN	0.9953	1.0000	–	0.9993	1.0000	1.0000	1.0000
LR	0.3784	1.0000	0.0007	–	1.0000	1.0000	1.0000
GBC	0.0000	0.7982	0.0000	0.0000	–	0.3340	0.5794
RF	0.0000	0.9386	0.0000	0.0000	0.6660	–	0.7654
E (All)	0.0000	0.8302	0.0000	0.0000	0.4206	0.2346	–

B   Supplemental Graphical Material

See Figs. 2, 3, 4, 5, 6, 7, 8 and 9

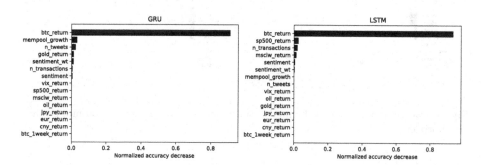

Fig. 2. Feature importance of the models with memory function on the 1-min prediction horizon.

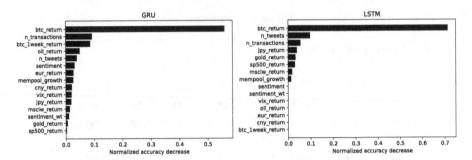

Fig. 3. Feature importance of the models with memory function on the 5-min prediction horizon.

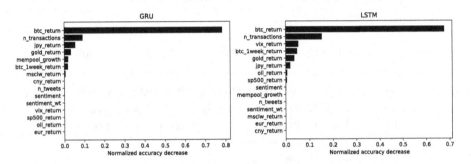

Fig. 4. Feature importance of the models with memory function on the 15-min prediction horizon.

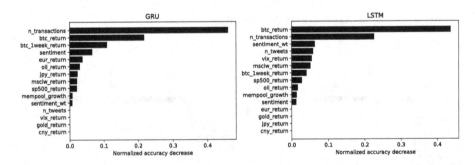

Fig. 5. Feature importance of the models with memory function on the 60-min prediction horizon.

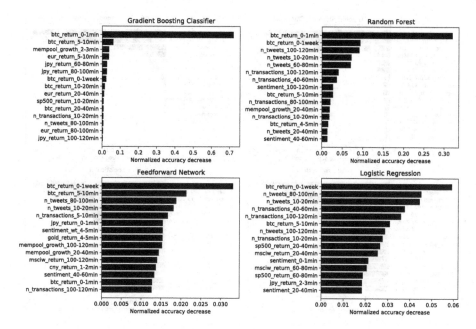

Fig. 6. Feature importance of the models without memory function on the 1-min prediction horizon.

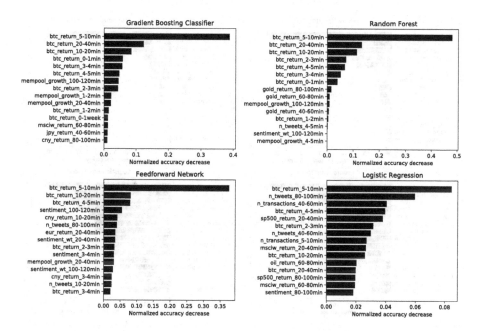

Fig. 7. Feature importance of the models without memory function on the 5-min prediction horizon.

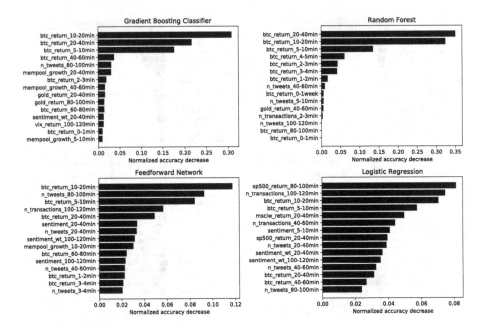

Fig. 8. Feature importance of the models without memory function on the 15-min prediction horizon.

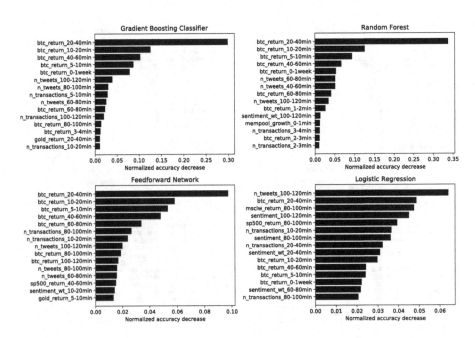

Fig. 9. Feature importance of the models without memory function on the 60-min prediction horizon.

References

1. Nakamoto, S.: Bitcoin: a peer-to-peer electronic cash system. Working paper (2008)
2. Feng, G., Giglio, S., Xiu, D.: Taming the factor zoo: a test of new factors. J. Finance **75**(3), 1327–1370 (2020)
3. Jaquart, P., Dann, D., Martin, C.: Machine learning for bitcoin pricing – a structured literature review. In: Proceedings of 15th International Business Informatics Congress, pp. 174–188 (2020)
4. Coinmarketcap: Coinmarketcap (2020). https://coinmarketcap.com/. Accessed 30 July 2020
5. Fama, E.F.: Efficient capital markets: a review of theory and empirical work. J. Finance **25**(2), 383–417 (1970)
6. Lo, A.W.: The adaptive markets hypothesis. J. Portf. Manag. **30**(5), 15–29 (2004)
7. Fama, E.F., French, K.R.: Dissecting anomalies. J. Finance **63**(4), 1653–1678 (2008)
8. Fischer, T., Krauss, C.: Deep learning with long short-term memory networks for financial market predictions. Eur. J. Oper. Res. **270**(2), 654–669 (2018)
9. Gu, S., Kelly, B., Xiu, D.: Empirical asset pricing via machine learning. Rev. Financ. Stud. **33**(5), 2223–2273 (2020)
10. Krollner, B., Vanstone, B., Finnie, G.: Financial time series forecasting with machine learning techniques: a survey. In: Proceedings of European Symposium on Artificial Neural Networks: Computational and Machine Learning, pp. 1–7. Springer (2010)
11. Fama, E.F.: Market efficiency, long-term returns, and behavioral finance. J. Financ. Econ. **49**(3), 283–306 (1998)
12. Grossman, S.J., Stiglitz, J.E.: On the impossibility of informationally efficient markets. Am. Econ. Rev. **70**(3), 393–408 (1980)
13. Green, J., Hand, J.R.M., Zhang, X.F.: The supraview of return predictive signals. Rev. Account. Stud. **18**(3), 692–730 (2013)
14. Schwert, G.W.: Anomalies and market efficiency. In: Finantial Markets and Asset Pricing, Handbook of the Economics of Finance, pp. 939–974. Elsevier (2003)
15. Dyhrberg, A.H.: Bitcoin, gold and the dollar – a GARCH volatility analysis. Financ. Res. Lett. **16**(1), 85–92 (2016)
16. Burniske, C., White, A.: Bitcoin: ringing the bell for a new asset class, Technical report (2017)
17. Urquhart, A.: The inefficiency of Bitcoin. Econ. Lett. **148**(1), 80–82 (2016)
18. Nadarajah, S., Chu, J.: On the inefficiency of Bitcoin. Econ. Lett. **150**(1), 6–9 (2017)
19. Bariviera, A.F.: The inefficiency of Bitcoin revisited: a dynamic approach. Econ. Lett. **161**(1), 1–4 (2017)
20. Vidal-Tomás, D., Ibañez, A.: Semi-strong efficiency of Bitcoin. Finance Res. Lett. **27**(1), 259–265 (2018)
21. Khuntia, S., Pattanayak, J.K.: Adaptive market hypothesis and evolving predictability of bitcoin. Econ. Lett. **167**(1), 26–28 (2018)
22. Karakoyun, E.S., Cibikdiken, A.O.: Comparison of ARIMA time series model and LSTM deep learning algorithm for bitcoin price forecasting. In: Proceedings of the Multidisciplinary Academic Conference, pp. 171–180 (2018)
23. McNally, S., Roche, J., Caton, S.: Predicting the price of bitcoin using machine learning. In: Proceedings of 2018 Euromicro International Conference on Parallel, Distributed, and Network-Based Processing, pp. 339–343 (2018)

24. Madan, I., Saluja, S., Zhao, A.: Automated bitcoin trading via machine learning algorithms (2015)
25. Smuts, N.: What drives cryptocurrency prices?: an investigation of google trends and telegram sentiment. ACM SIGMETRICS Perform. Eval. Rev **46**(3), 131–134 (2019)
26. Goodfellow, I., Bengio, Y., Courville, A.: Deep Learning. MIT Press, Cambridge (2016)
27. Kumar, A., Garg, G.: Sentiment analysis of multimodal Twitter data. Multimed. Tools Appl. **78**(17), 24103–24119 (2019)
28. Symeonidis, S., Effrosynidis, D., Arampatzis, A.: A comparative evaluation of pre-processing techniques and their interactions for twitter sentiment analysis. Expert. Syst. Appl. **110**(1), 298–310 (2018)
29. Takeuchi, L., Lee, Y.-Y.A.: Applying deep learning to enhance momentum trading strategies in stocks (2013)
30. Kingma, D.P., Ba, J.: Adam: a method for stochastic optimization. In: Proceedings of 3rd International Conference on Learning Representations, pp. 1–15 (2015)
31. Hornik, K., Stinchcombe, M., White, H.: Multilayer feedforward networks are universal approximators. Neural Netw. **2**(5), 359–366 (1989)
32. Hochreiter, S., Schmidhuber, J.: Long short-term memory. Neural Comput. **9**(8), 1735–1780 (1997)
33. Chung, J., Gülçehre, Ç., Cho, K., Bengio, Y.: Empirical evaluation of gated recurrent neural networks on sequence modeling. Working Paper (2014)
34. Ho, T.K.: Random decision forests. In: Proceedings of 3rd International Conference on Document Analysis and Recognition, pp. 278–282 (1995)
35. Friedman, J.H.: Greedy function approximation: a gradient boosting machine. Ann. Stat. **5**(21), 1189–1232 (2001)
36. Diebold, F.X., Mariano, R.S.: Comparing predictive accuracy. J. Bus. Econ. Stat. **20**(1), 134–144 (1995)
37. Breiman, L.: Random forests. Mach. Learn. **45**(1), 5–32 (2001)
38. Fischer, T.G., Krauss, C., Deinert, A.: Statistical arbitrage in cryptocurrency markets. J. Risk Financ. Manag. **12**(1), 31 (2019)
39. Biais, B., Bisiere, C., Bouvard, M., Casamatta, C., Menkveld, A.J.: Equilibrium bitcoin pricing. Working Paper (2018)
40. Jegadeesh, N.: Evidence of predictable behavior of security returns. J. Finance **45**(3), 881–898 (1990)
41. Shefrin, H., Statman, M.: The disposition to sell winners too early and ride losers too long: theory and evidence. J. Finance **40**(3), 777–790 (1985)

Fraud Detection and Information Generation in Finance

Scalable and Imbalance-Resistant Machine Learning Models for Anti-money Laundering: A Two-Layered Approach

Pavlo Tertychnyi[1(✉)], Ivan Slobozhan[1], Madis Ollikainen[2], and Marlon Dumas[1]

[1] University of Tartu, Tartu, Estonia
pavel.tertychny@gmail.com, ivan.slobozhan@gmail.com, marlon.dumas@ut.ee
[2] Tallinn University of Technology, Tallinn, Estonia
madisollikainen@gmail.com

Abstract. In this paper, we address the problem of detecting potentially illicit behavior in the context of Anti-Money Laundering (AML). We specifically address two requirements that arise when training machine learning models for AML: scalability and imbalance-resistance. By scalability we mean the ability to train the models to very large transaction datasets. By imbalance-resistance we mean the ability for the model to achieve suitable accuracy despite high class imbalance, i.e. the low number of instances of potentially illicit behavior relative to a large number of features that may characterize potentially illicit behavior. We propose a two-layered modelling concept. The first layer consists of a Logistic Regression model with simple features, which can be computed with low overhead. These features capture customer profiles as well as global aggregates of transaction histories. This layer filters out a proportion of customers whose activity patterns can be deemed non-illicit with high confidence. In the second layer, a gradient boosting model with complex features is used so as to classify the remaining customers. We anticipate that this two-layered approach achieves the stated requirements. Firstly, feature extraction is more scalable as the more computationally demanding features of the second layer do not need to be extracted for every customer. Secondly, the first layer acts as an undersampling method for the second layer, thus partially addressing the class imbalance. We validate the approach using a real dataset of customer profiles and transaction histories, together with labels provided by AML experts.

Keywords: Machine learning · Anti-money laundering

1 Introduction

Money Laundering (ML) is a sophisticated illegal activity where a group of economic actors collaborate to obscure the origin and real beneficiaries of monetary funds and transactions. Governmental authorities fight against ML and Terrorist Financing (TF), in part by setting regulations on financial institutions.

B. Clapham and J.-A. Koch (Eds.): FinanceCom 2020, LNBIP 401, pp. 43–58, 2020.
https://doi.org/10.1007/978-3-030-64466-6_3

Those regulations force banks to detect potentially illicit activity in their customers' accounts that might be associated with ML or TF. As the number of transactions per minute may be in the order of millions, reviewing every transaction manually is impractical. Accordingly, banks rely on automated monitoring systems to sift through transactions and raise alerts in case of potentially illicit activity.

Traditional monitoring systems for AML (Anti-Money Laundering) are rule-based [1]. These systems generate a large proportion of false alerts, which in turn creates a high workload for the AML experts. According to McKinsey & Co [2] this drawback stems from the fact that rule-based systems are not able to capture complex relationships. Given the availability of vast volumes of transaction data in banking systems, there is a fertile ground for complementing rule-based systems with machine learning algorithms [3,4].

The development of machine learning models for ML detection raises a tension between scalability and accuracy. AML monitoring approaches based on Logistic Regression (LR) or Decision Trees (DT) cannot effectively capture the complex patterns typically found in ML, while approaches based on Support Vector Machines (SVM), Neural Networks (NN) or deep learning are computationally expensive. A typical financial institution handles millions of customer accounts and hundreds of millions of transactions. Tradeoff solutions are tree-based techniques such as Random Forests (RF) and Gradient Boosting Trees (GBT) which have been shown to achieve high levels of accuracy for a wide range of classification tasks in various domains [5], while providing suitable levels of computational efficiency. However, the computational efficiency of such methods highly depends on the complexity of calculating the feature set.

In this setting, we address the question of how to train accurate machine learning models for AML, while fulfilling two requirements: scalability and imbalance-resistance. By scalability we mean the ability to train the models to very large transaction datasets, in such a way that retraining the model periodically with relatively standard computing resources is feasible. This requirement precludes approaches that entail generating sophisticated feature sets for every customer or training deep learning models on the entire dataset. By imbalance-resistance we refer to the ability for the model to achieve suitable accuracy despite high class imbalance, i.e. the low number of instances of potentially illicit behavior relative to a large number of features that may characterize potentially illicit behavior. To improve imbalance-resistance, machine learning models use well-known approaches to handle class imbalance via under-sampling or over-sampling. However, in our context, this random sampling approach does not exploit the fact that a very large proportion of customers have transactional patterns that make them unlikely to engage in illicit behavior.

To address the above requirements, we propose a two-layered architecture for training machine learning classifiers for detecting potentially illicit financial behavior. In this architecture, two classifiers are applied sequentially to the input samples. The first classifier relies on a highly efficient LR classification technique coupled with a small number of simple features capturing the customer profile and aggregate transaction volumes. This layer is intended to filter out clearly non-illicit customers. The second layer then uses a larger and more sophisticated set of features and a more complex classifier (extreme gradient boosting) in order to classify heightened risk customers into potentially illicit and non-illicit. The proposed framework incorporates a range of approaches for feature extraction, including mean-encoded categorical features, statistics based on an aggregation of time series data, customer ego-network statistics, and features extracted from stochastic models.

The framework has been evaluated on a real-life large-scale dataset consisting of customer profiles and transaction histories, together with labels provided by AML experts within a universal bank.

The rest of the paper is organized as follows. Section 2 briefly reviews the relevant literature. Section 3 depicts the data we used for model development. Section 4 describes the methodology including layered model definition, feature extraction and tackling the dataset imbalance. Section 5 shows the experimental setup we designed. Section 6 shows the results of the experiments. Section 7 opens the discussion of machine learning model limitations, sketches the directions of the future research and summarizes our main conclusions and contributions.

2 Related Work

The application of data analytics techniques for financial fraud and ML detection goes back to the 1990s [6]. A notable pioneer is the FAIS expert system [7], developed in the Financial Crimes Enforcement Network (the US Financial Intelligence Unit). FAIS relies on a manually maintained set of rules to identify parties with potentially illicit transactional patterns. According to Chen et al. [8] such rule-based AML systems have dominated industrial standards for decades. Rule-based systems have two strong advantages: they are easy to interpret and can be designed by domain experts with a little technical background. However, they offer limited flexibility, are cumbersome to update and adjust to changing environment and rely solely on expert knowledge of criminal behavior [8]. Rule-based monitoring in combination with behavior detection monitoring and link analysis based monitoring was used as a framework by Helmy et al. [9] to tackle ML. In addition to it, the authors applied clustering techniques to reduce false alerts raised by the framework. Stepping ahead, some researches have moved away from strict rules and studied fuzzy rules for detection of ML behavior. One of the examples is a work of Chen et al. [10]. The solution has been tested on 710 accounts from Google File System.

Accordingly, modern literature has been moving towards statistical and machine learning approaches for AML/CTF potentially illicit activity detection [8]. A wide range of approaches has been investigated in this field, including linear models and tree-based techniques.

Tree-based machine learning techniques on their own and in ensemble with other algorithms were studied in many types of research. S. Kotsiantis et al. [3] explored the problem of Fraudulent Financial Statements (FFS) using a combination of multiple machine learning methods including DT, NN, SVM and

LR. The developed ensemble method was trained and tested on a dataset of 164 Greek manufacturing firms, where for 41 of them there was published an indication of potentially illicit behavior. The resulting model achieved reasonably good accuracy varying from 91.2 to 95.1 depending on an ensembling approach. Another example of industry solution is present in a work of V. Jayasree et al. [4] where authors have evaluated the regulatory risk of ML using Bitmap Index-based DT (BIDT). Their solution was developed based on Statlog German Credit Data and shown better results in terms of risk identification time, FPR and TPR than existing by that time Smart Card-based Security Framework [11] and Multilayered Detection System [12].

Less research is available for application of linear models for AML detection. Liou et al. [13] compared LR, NN and DT on the Taiwan Economic Journal data bank [14]. All three methods showed impressive results with accuracy in a range from 95.59% to 99.05% on a balanced dataset of 6037 firms total.

To our knowledge and available literature, there is no published research on the usage of boosting and begging DT ensemble methods such as RF and GBT. In addition to this, there is a lack of systematic exploration of different feature extraction approaches and their relative performance. Instead, each study uses an ad-hoc set of manually engineered features.

Last but not least, a bottleneck to train machine learning models is the availability of data, given that financial transaction data is highly sensitive. Some of the solutions are developed and tested on artificially created data. Lopez-Rojas et al. [15] tested Rule-based models, DTs and clustering techniques on financial data synthesized by Multi-Agent-Based Simulation approach. In this study, we rely exclusively on real-life data extracted from banking systems across three countries and covering both private and corporate sectors.

3   Data

This work has been made in collaboration with a universal bank, which has made available data, where all personal identifiers have been replaced with non-informative pseudonyms, to be processed within its premises and infrastructure. In the following, we define the labels used for supervised learning and provide a sketc.h of the database structure used.

We use as labels information about instances of customer activities that AML experts have previously deemed to as having reasonable grounds for reporting to the local Financial Intelligence Unit (FIU). Figure 1 illustrates a generalized AML alert processing workflow. Activities of a customer can be flagged by different types of detection engines, e.g. automatic monitoring tools, escalations from bank's branch offices or even by tips from external sources. The flagged activities could, for example, be single transactions or sets of transactions over some period. The bank's AML experts manually analyse all the flagged activities in the wider context of the customers behaviour, history, counterparties, etc. Some alerts are immediately classified as non-illicit while others require deeper investigation; for example, an AML expert may ask for additional documents and explanations from a customer. Finally, when an investigation is done the AML expert decides whether there are reasonable grounds for filing a report to the FIU. In our research we use two labels: customers who have been previously reported to the FIU and those who have not. For the later we use a random sample from the customer base, thus in the following they are called *Randomly Sampled* customers.

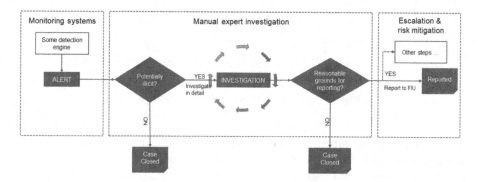

Fig. 1. Alert processing workflow.

Figure 2 gives a sketch of the database structure, which is used for the present research. It consists of four tables: customers' identifiers table, demographic data table, transactions data table, reports table. In demographic data table, there is static information which describes customers, while in the transactions data table, there are transactions of those customers. A transaction has several properties like a direction – whether the transaction is incoming or outgoing, activity type – whether the transaction is foreign or intrabank, etc., channel – whether the transaction was made through physical or electronic channels, etc., and others which are self-explanatory by their names. Reports table has information about reports on those customers who deemed to be involved in ML activity.

Obviously, potentially illicit activity is uncommon, thus we have high class imbalance in the present research. The detailed explanations of how we tackled the class imbalance could be found in Sect. 4.4. ML is a complex activity and with high likelihood involve a longer period of activity. Thus, we have decided to approach the problem on longer term customer activity level, not on single transaction level. Accordingly, the labels used are on a customer activity period level, not on transaction level. Note, that for banks from different countries, the criteria for reporting to the FIU could be different due to differences in legislation. The possible implications from this are discussed in Subsect. 6.2.

4   Methodology

In this section, we describe in details the framework we developed to detect potentially illicit financial behavior. First, we define the framework with an

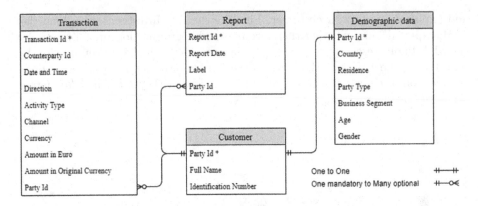

Fig. 2. The sketch of a database structure.

explanation on its architecture and reasons which lead to it. Then, we are going to describe types of features we generated based on raw demographical and transactional customers' data together with the reasoning of generating each type of those features. In the end, we illustrate how we tackled the imbalance of data.

4.1 Two-Layered Model Concept

Obviously, the overwhelming majority of customers are clearly non-illicit, and only a small portion of a customer base is worth to be investigated. If we filter out such clearly non-illicit customers, we can significantly narrow down the problem. From another point of view, this filtering solves some part of the computational cost problem, since it reduces the number of customers for whom we need to extract complex features. For these purposes, we introduce a two-layered model concept. First layer model is fast and simple but less accurate. It is focused on dropping "obviously non-illicit" customers. In turn, the second layer is using a more robust classifier trained on customers represented with a larger set of complex features. The aim of the second layer model is to distinguish customers who worth to be checked by AML experts among all heightened risk customers. On Fig. 3 you may find the architecture of the final classification model.

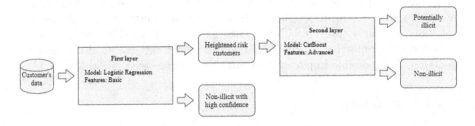

Fig. 3. Final classification model.

As a first layer, we use a linear model trained on a set of simple features: standard deviation, mean, and count of transactions; business segment type, customer type; simple statistic of an egocentric network of each customer. These features are fast to compute, and they clearly describe customers with small turnover. As a first layer classifier we use Logistic Regression because of its simplicity and speed of training. As the second layer classifier, we use extreme gradient boosting trees, specifically the CatBoost implementation of this latter algorithm [16], which is tailored to handle mixtures of numerical and categorical features.[1] In this research, the Randomly Sampled customers were used as a negative class and Reported customers as a positive class.

4.2   Framework

All techniques we used are combined into a monitoring framework (see Fig. 4). As input data, we used K months snapshots of customers' activity (see Sect. 5 for details). Then for each of the layers we extract own feature sets. We do not perform any feature selection for the first layer. Parameters of the first layer model are tuned by Grid search of parameters. For the second layer we do feature selection and we use Bayesian Optimization for hyperparameter tuning. For the first layer we use fast Logistic Regression model and for the second layer we use the CatBoost model. Customers for whom the first-layer model produces a class probability above a predefined threshold, form the customer base for the second layer (heightened risk customers). This first-layer threshold is tuned to minimize false negatives as explained later (see Sect. 5). The second layer takes as input K months snapshots of heightened risk customers. Alerts are generated to customers for whom the second layer produces a class probability above a threshold, which is determined by the investigators.

4.3   Feature Extraction

In our approach, we focus on customer-level prediction while having two sources of raw data: demographical and transactional. Demographical data is static and customer-based, therefore it does not require complex feature engineering. The main issue arises in utilizing dynamic time-series data of customers' transactions. Below, you may find a table (see Table 1) with a description of different types of features we generated for the classifier, reasoning behind them and correspondent examples. All features were calculated by standard groupby-apply operations, except of Sequence-based features. For them the more detailed calculation is provided in the Appendix.

Eventually, we have more than 400 features for the second layer, and it is too inefficient to extract all of them, thus we have to do a robust feature selection. All features are firstly filtered through high correlation and low variance filters.

---

[1] We also conducted experiments using another implementation of extreme gradient boosting (XGBoost), but these experiments consistently led to lower accuracy. In the evaluation reported below, we only report on results obtained using Catboost.

Table 1. Features types with their reasoning and examples.

Name	Definition	Reasoning	Example
Demographical Features	Static demographical characteristics of customers	Potentially illicit patterns vary significantly depending on the type of a party (business or private), country, business segment, etc.	Business segment, country of residence, SIC code
Simple Descriptive Statistics	Descriptive statistics of transactions amount (min, max, mean, std, count, sum, median) aggregated by various categorical columns: by direction, by channel or through the whole data range	Based on an exploratory analysis of the data, we discovered that even simple statistical measures have different distributions for potentially illicit and non-illicit customers	Mean amount of the inbound transactions a customer made during the last month
Transaction Time Difference Features	Statistics that calculated based on the difference between timestamps of two consecutive transactions	An observation was that in some cases, potentially illicit customers made very frequent transactions over a small period of time	Min, max, mean time between consecutive transactions
Transaction Amount Difference Features	Differences between sums and counts of aggregations for different time periods	Potentially illicit behavior may drastically change over some periods of time. For example, in one month criminals might transact very frequently and with large volumes of money, but after that, there can be a quite a long period of lull	The difference between the sum of incoming transactions for September and October, the difference between the sum of incoming versus outgoing transactions for August
Counterparty Features	Statistics based on information from counterparties for those customers, who have at least one transaction with a defined counterparty	Since ML often is a group activity we need to have some characteristics of customers' counterparties	Ratios between unique counterparty types and amount of transactions, the number of unique counterparty types (i.e. accounts, names, etc.)
Proportion Features	Proportions of channels and activity types, aggregated by time and direction	The distribution of customers' transactions may be different for potentially illicit than for non-illicit customers, we assume that they can have some abnormalities in transactions type and channels	Proportion of Payments in October from all transactions in October, a proportion of cash withdrawals in June from all transactions in June
Sequence-based Features	Generative log-odds features [17], where we estimate transition probabilities between each transaction state separately for potentially illicit and non-illicit customers and then compare them. The feature is calculated as $\log \frac{P(Y = potentially\ illicit \mid x_1, \ldots, x_n)}{P(Y = non-illicit \mid x_1, \ldots, x_n)}$ where $x_1, x_2, \ldots, x_n$ are some discrete properties of transactions (i.e. direction). Probabilities in numerator and denominator are calculated using Bayes theorem and probabilities of sequences estimated with the chain rule and simplified assumptions of Markov property	Sequences of customers' transactions are not random and should follow some hidden structure	Log-odds feature where the sequence is direction, channel, activity type, discretized transaction amount

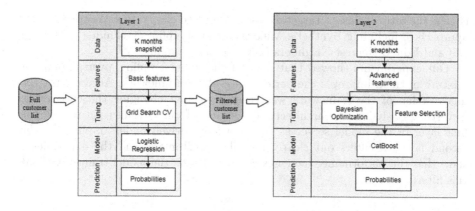

Fig. 4. Proposed framework for AML monitoring.

After, we apply BoostARoota algorithm - an extension of Boruta algorithm [20] for all relevant feature selection. In this feature selection technique importance of each feature is compared to importance of its "shuffled version". If a feature is more important than random, with a predefined confidence level, then it is kept, otherwise not. We created an automated features selection method that allows us to automatically choose the most relevant features and reduce overfitting. The features are selected on a validation set with respect to the best Area Under the Precision-Recall Curve (AUPRC). In all cases, applying BoostARoota algorithm we reduced approximately 25% of features and gained marginal improvement in AUPRC. At the same time, by reducing the number of features, Catboost model is trained faster, allowing us to utilize deeper trees within a reasonable time.

4.4 Tackling Class Imbalance

As it was described earlier, we are dealing with a highly imbalanced dataset. In order to build a meaningful model and correctly estimate its performance, we need to pay attention to class ratio. In our approach, we used two common ways of dealing with imbalanced datasets: cost-sensitive learning and sampling techniques. Later we discuss the metrics we used to estimate model performance.

Cost-Sensitive Learning. Cost-sensitive learning is a common approach to deal with class imbalance by penalizing a model more for incorrectly predicting the minority class versus the majority class. In this approach, we add weights for each class in a cost function, therefore the more weight a particular class has, the higher loss will be given for an incorrectly classified example. In our study, we wish to penalize model more for incorrect prediction of the Reported case, rather than for Randomly Sampled case.

Combination of Undersampling and Oversampling. There are exist different ways of dataset sampling: undersampling of a majority class, oversampling of a minority class and their hybrid approaches. In all cases, we artificially change the ratio of classes, targeting less imbalanced class ratios. We tried several approaches, including Synthetic Minority Over-sampling Technique (SMOTE) [18] and Edited Nearest Neighbors (ENN) under-sampling [19].

Our experiments showed that none of the undersampling, oversampling or mixture of them improves the performance of the framework. Accordingly, the final version of the framework was trained without changing the class ratio. Nevertheless, due to the architectural decision of our classification model, the first layer of the model serves as the knowledge-driven undersampling for the second layer. It filters out clearly non-illicit customers, and the experiments show that the performance increases compared with the model trained without this filtering.

5   Experimental Setup

In the experiments, we use a subset of Reported and subset of Randomly Sampled bank's customers. Total bank's database is in order of magnitude of millions, but for experimental purposes, we use ∼330000 customers from 3 countries with more than 51 million transactions. The percentage-wise distribution is 0.004% of Reported and 0.996% of Randomly Sampled customers; 8% of corporate and 92% of private customers. All customers have a transaction history from August 2017 to December 2018.

As samples for classification, we take snapshots of customers at a given period of time, and for each customer we take K months history. In this study, we set K equals to 6. For the Reported customers – K months prior to the date of AML experts' decision, for Randomly Sampled customers – K random consequent months. For clarity, the schematic explanation of snapshots extraction is presented on a Fig. 5.

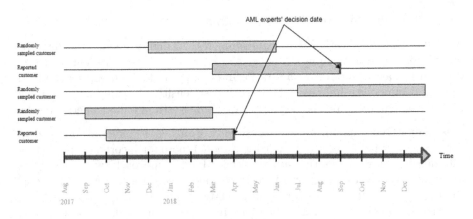

Fig. 5. Example of snapshots creation.

Training schema for the framework is inspired by the training schema of stacked models, but in our case, the second layer model doesn't use the output of the first layer directly as features. Instead, it uses a richer feature set than the first layer and it is trained and tested only on prefiltered customers. Data split on train and test is stratified by country, party type and label.

Since the dataset is highly imbalanced, we use the AUPRC as a measure of accuracy. Both model layers are trained to maximize AUPRC separately because they have different thresholds, purposes, and constraints. The purpose of the first layer is to filter out clearly non-illicit customers and at the same time do not miss potentially illicit ones, thus, the probability threshold is set in such a way that the first layer does not miss more than 1% of potentially illicit customers on a validation set. In turn, the probability threshold for the second layer could be tuned by AML or compliance officers in a risk-based approach in accordance to their institutes risk appetite. To illustrate the options for risk-based tuning we also use precision and recall at fixed thresholds. Both of those metrics are crucial since precision defines how many alerts made are correct and recall defines what is the proportion of potentially illicit customers for whom alerts are generated.

6   Results

As stated in Sect. 1, we anticipate that the proposed two-layered model architecture outperforms a single-layered one. To validate it, we conducted an experiment where we compare the performance of the two-layered model with baseline RF and CatBoost models. Also, we measured the execution time of full model training and application pipelines for all three models.

6.1   First Layer Results

Firstly, we apply the first layer model on the full customer set for filtering. The decision threshold is chosen in such a way that with this threshold we filter out no more than 1% of Reported customers on the validation set. Thereby, on a test set, we filter out 48% of Randomly Sampled customers and miss 6.6% of the Reported customers. Below is the confusion matrix for the first layer (Table 2).

Table 2. Confusion matrix for the first layer.

		Predicted	
		Randomly sampled	Reported
Actual	Randomly sampled	48.07%	51.93%
	Reported	6.6%	93.4%

6.2   Second Layer Results

Table 3 shows the observed performance for the second layer of the model for each of the countries. As we can see, the results are different from country to country. The classifier has the best performance for Country 1. One of the potential reasons is that the labels for three different countries are created by different groups of AML experts, and it is possible that the criteria used by them to report customers in Country 1 is better recognized by our features than in Country 2 and Country 3. Another potential reason is the dissimilarity in legislation. Banks from different countries have different reporting requirements and guidelines from respective authorities. Accordingly they might have different sets of rules by which they raise alerts. Some of those rules are unique and ad hoc, so it creates the difference in distributions of labels for different countries.

Table 3. Performance of the second layer on three countries separately.

Metric	Threshold 0.25		Threshold 0.5		Threshold 0.75		
	Recall	Precision	Recall	Precision	Recall	Precision	AUPRC
Country 1	0.859	0.321	0.703	0.576	0.546	0.813	0.732
Country 2	0.627	0.412	0.450	0.575	0.267	0.759	0.517
Country 3	0.582	0.291	0.373	0.472	0.208	0.730	0.422

We also observed even stronger decrease in performance for all three countries if we train three different models for each country separately. This proves that fraudsters in all countries share some patterns, and it is crucial to train one model for the whole dataset, even though the way customers are labelled is slightly different for the countries.

6.3   Overall Results

Below we present the overall performance of the two-layered model. From the first layer, we take customers who have been classified as "non-illicit with high confidence" (TN and FN from the confusion matrix in Table 2) and add those to "heightened risk" customers classified by the second layer. To get final confusion matrices, we have to sum up TNs and FNs from both layers and take TPs and FPs only from the second layer.

To measure the performance of the two-layered model, we compared it to baseline models – RF model and CatBoost model which were trained and tested on the not filtered customer base. As features, we used the complex feature set which in the two-layer model we used only for the second layer. The results for all countries together can be found in Table 4.

We also measured the execution time of major stages of model training. The execution times are shown in Table 5. We note that the two-layered model is faster to train than the single-layer RF and CatBoost models, since the two-layered architecture allows us not to extract complex features for all customers, which is the most time-consuming stage. Also, we notice that the CatBoost model is slightly slower to train that the RF model.

Table 4. Performance on three countries together.

| | Threshold 0.25 | | Threshold 0.5 | | Threshold 0.75 | | |
Metric	Recall	Precision	Recall	Precision	Recall	Precision	AUPRC
Two-layer model	0.618	0.348	0.448	0.548	0.287	0.772	0.497
Catboost model	0.615	0.298	0.430	0.531	0.209	0.775	0.467
RF model	0.615	0.218	0.439	0.447	0.3	0.622	0.428

We can observe that the two-layered model outperforms the CatBoost model and RF model by both the time of training and performance. Also, we can notice that the Random Forest model is beaten by the CatBoost model in performance.

All experiments were run on a PC with Intel Core i5-6300U CPU (x64-based 2.40 GHz processor), 64 GB RAM, with a 64-bit Operating System.

Table 5. Execution time by stages.

	Train	Application
Number of samples	~230k	~100k
Stages included	Samples feature extraction + model(s) training	Samples feature extraction + model(s) application
Two-layer model time	89 min	42 min
Catboost model time	148 min	66 min
RF model time	143 min	65 min

7   Conclusions and Future Work

In this paper, we presented and evaluated a two-layered approach to train machine learning models for detecting potentially illicit behavior in the context of AML monitoring. The key idea developed in the paper is that instead of training a single classifier using an extensive set of features, we can train a first simple model to discard "clearly non-illicit customers" and pipeline it with a second model that uses a more sophisticated feature set and classifier learning method. We instantiated this general two-layered concept using LR for the first layer, and extreme gradient boosting trees Catboost methods for the second layer. The first layer relies on "static" customer-level features and transaction volume aggregates, while the second layer combines several feature sets, ranging from aggregates for different time windows and attributes to Markov Chain based stochastic models to capture the temporal dynamics of transactions. We advocated that this architecture allows us to achieve better scalability while allowing us to handle the high class imbalance typically found in the field of AML.

From the experiments we made, we found out that the two-layered approach outperforms a single-layered approach based on a single classifier (CatBoost and RF) both in terms of accuracy and execution time for model training. Also, we showed that CatBoost model outperforms the RF model in terms of accuracy. We also discovered that training a single model to make predictions across multiple countries leads to higher accuracy than training separate models for each country.

A fundamental limitation of the proposed approach is that it is a purely supervised learning solution, which means that the framework is able to detect only those illicit behavior patterns that have been previously identified by the AML experts who provide the labels. In the reported experiments, we observed that the information encoded in the labelled data could bias the model significantly. A possible direction to enhance the proposed approach would be to combine it with anomaly detection methods, so the method will help to identify new potentially illicit behavior together with the behavior that investigators have previously labelled as potentially illicit.

Another limitation of the approach is that it focuses on extracting features from tabular data. In the case when AML experts are not certain regarding the status of a customer, they may ask for additional documents, such as invoices, contracts, and purchase orders, etc. Exploiting this information, via text mining techniques, is another possible extension.

Finally, the proposed study focused on calculating the probability that a given customer (at a given point in time) engages in behavior that may be considered to be potentially illicit. We did not address the question of when to trigger an alert (i.e. trigger an investigation) based on the generated predictions. When designing an alerting system in this context, one needs to take into account the availability of resources to conduct investigations, and the fact that generating alerts earlier is generally preferable than later, but generating an alert too early may lead to there not being sufficient information to conduct an investigation. Researching this question is an avenue for future work.

Acknowledgements. This research was partly funded by the European Regional Development Funds via Archimedes Foundation (NUTIKAS programme).

A   Appendix

Sequence-based Features Calculation. We have an assumption that sequences of customers' transactions are not random and follow some hidden structure. Therefore we used a way to encode this information to the model by so-called generative log-odds features [17], where we estimate transaction probabilities between each transaction state separately for potentially illicit and non-illicit customers and then compare them. This approach allows us to capture the dynamics of the transaction history for our classification task while introducing less overhead than methods based on neural networks (e.g. Boltzman machines) or deep learning auto-encoders. In the log-odd feature extraction method, we want to generate features based on sequential probabilities. We are interested in the following probability:

$$P(X) = P(x_1, x_2, \ldots, x_n) \tag{1}$$

where $x_1, x_2, \ldots, x_n$ are some discrete properties of transactions (i.e. direction). One particular way to estimate this probability is to use the chain rule:

$$P(X) = P(x_1, \ldots, x_n) = p(x_1)p(x_2 \mid x_1) \ldots p(x_n \mid x_1, \ldots, x_{n-1}) \tag{2}$$

In some cases, it is practically impossible, so we can simplify the assumptions using Markov property:

$$P(X_n = x_n \mid X_{n-1} = x_{n-1}, \ldots, X_0 = x_0) = P(X_n = x_n \mid X_{n-1} = x_{n-1}) \tag{3}$$

But for our task, we are more interested in finding that a particular set of transactions is more illicit than just a set of regular non-illicit transactions. Mathematically, we want to estimate:

$$\text{argmax}_{y \in (potentially\ illicit, non-illicit)} P(Y = y|X) \tag{4}$$

One way to calculate this probability is to use Bayes theorem:

$$\underset{y}{\text{argmax}}\, P(Y = y \mid X) = \underset{y}{\text{argmax}}\, P(X \mid Y = y)P(Y = y) \tag{5}$$

The only thing left is to calculate $P(X \mid Y = y)$ and $P(Y = y)$. $P(X \mid Y = y)$ can be calculated using train set and then calculating transition probabilities separately for potentially illicit class and non-illicit class. For example, if there are only two states in a transaction sequence, namely in, out. All we need to estimate transition probabilities is to calculate

$$P(in \mid out) = \frac{(count(out \to in))}{(count(out))} \tag{6}$$

Similarly, for other combination of in, out we should do the same. $P(Y = y)$ is the prior probability of being potentially illicit, which is simply a proportion of potentially illicit customers in a full customer set for train data. Finally, instead of outputting a binary label 1/0 (potentially illicit sequence or not), we can plug in this as a feature into a classifier along with other features. We can use so-called log-odds ratio instead of a binary feature, defining as:

$$\cdot \log \frac{P(Y = potentially\ illicit \mid X)}{P(Y = non - illicit \mid X)} \tag{7}$$

References

1. Tsui, E., Gao, S., Xu, D., Wang, H., Green, P.: Knowledge-based anti-money laundering: a software agent bank application. J. Knowl. Manage. (2009)

2. Breslow, S., Hagstroem, M., Mikkelsen, D., Robu, K. The new frontier in anti-money laundering McKinsey Insights, November 2017. https://www. mckinsey.com/business-functions/risk/our-insights/the-new-frontier-in-anti-money-laundering

3. Kotsiantis, S., Koumanakos, E., Tzelepis, D., Tampakas, V.: Forecasting fraudulent financial statements using data mining. Int. J. Comput. Intell. **3**(2), 104–110 (2006)

4. Jayasree, V., Siva Balan, R.V.: Money laundering regulatory risk evaluation using bitmap index-based decision tree. J. Assoc. Arab Univ. Basic Appl. Sci. **23**(1), 96–102 (2017)

5. Nielsen, D.: Tree boosting with XGBoost - why does XGBoost win "every" machine learning competition? Master's Thesis, NTNU (2016)

6. Palshikar, G.K., Apte, M.: Financial Security Against Money Laundering: A Survey. In: Emerging Trends in ICT Security, pp. 577–590. Morgan Kaufmann (2014)

7. Senator, T.E., et al.: Financial crimes enforcement network AI system (FAIS) identifying potential money laundering from reports of large cash transactions. AI Mag. **16**(4), 21 (1995)

8. Chen, Z., Teoh, E.N., Nazir, A., Karuppiah, E.K., Lam, K.S.: Machine learning techniques for anti-money laundering (AML) solutions in potentially suspicious transaction detection: a review. Knowl. Inf. Syst. **57**(2), 245–285 (2018)

9. Helmy, T.H., Zaki, M., Salah, T., Badran, K.: Design of a monitor for detecting money laundering and terrorist financing. J. Theoret. Appl. Inf. Technol. **85**(3), 425 (2016)

10. Chen, Y.T., Mathe, J.: Fuzzy computing applications for anti-money laundering and distributed storage system load monitoring (2011)

11. Cortinas, R., et al.: Secure failure detection and consensus in trustedpals. IEEE Trans. Dependable Secure Comput. **9**(4), 610–625 (2012)

12. Phua, C., Smith-Miles, K., Lee, V., Gayler, R.: Resilient identity crime detection. IEEE Trans. Knowl. Data Eng. **24**(3), 533–546 (2010)

13. Liou, F.M.: Fraudulent financial reporting detection and business failure prediction models: a comparison. Manage. Audit. J. (2008)

14. tej.com.tw

15. Lopez-Rojas, E.A., Axelsson, S.: Money laundering detection using synthetic data. In: The 27th annual workshop of the Swedish Artificial Intelligence Society (SAIS), Örebro; Sweden, 14–15 May 2012, no. 071, pp. 33–40. Linköping University Electronic Press, May 2012

16. Prokhorenkova, L., Gusev, G., Vorobev, A., Dorogush, A.V., Gulin, A.: CatBoost: unbiased boosting with categorical features. In: Advances in Neural Information Processing Systems, pp. 6638–6648 (2018)

17. Leontjeva, A., Goldszmidt, M., Xie, Y., Yu, F., Abadi, M.: Early security classification of skype users via machine learning. In Proceedings of the 2013 ACM workshop on Artificial Intelligence and Security, pp. 35–44. ACM, November 2013

18. Chawla, N.V., Bowyer, K.W., Hall, L.O., Kegelmeyer, W.P.: SMOTE: synthetic minority over-sampling technique. J. Artif. Intell. Res. **16**, 321–357 (2002)

19. Wilson, D.L.: Asymptotic properties of nearest neighbor rules using edited data. IEEE Trans. Syst. Man Cybern. **3**, 408–421 (1972)

20. Kursa, M.B., Jankowski, A., Rudnicki, W.R.: Boruta-a system for feature selection. Fundamenta Informaticae **101**(4), 271–285 (2010)

Leveraging Textual Analyst Sentiment for Investment

Matthias Palmer[1]($\boxtimes$) and Timo Schäfer[2]

[1] University of Goettingen, Goettingen, Germany
matthias.palmer@uni-goettingen.de
[2] Goethe University Frankfurt, Frankfurt am Main, Germany
t.schaefer@wiwi.uni-frankfurt.de

Abstract. We document that the sentiment conveyed in texts of reports written by financial analysts is informative about both contemporaneous and future stock prices. By setting up a portfolio trading strategy exploiting textual sentiment in analyst reports, we show that it is possible to generate average monthly factor-adjusted returns of 0.7%. In this context, we find that the past price target forecasting abilities of brokerage firms have a positive effect on the predictability of returns using sentiment portfolio strategies. Overall, our results demonstrate that analysts provide valuable information for interpreting and predicting stock price movements in their textual reports. In contrast to existing research that utilizes quantitative analyst information or analyzes textual analyst data with a focus on event studies, we stand out by conducting an analysis at the calendar-year level to leverage on qualitative analyst report content. Most importantly, our results demonstrate that the financial sector still offers untapped potential for the inclusion of qualitative information that can be relevant for both research and practice.

Keywords: Financial analysts · Textual analysis · Portfolio strategy

1 Introduction

Texts in analyst reports help to better understand quantitative measures reported by financial analysts, e.g., stock recommendations, price targets, and earnings forecasts, and also provide interpretations of corporate earnings results as well as quarterly conference calls [1, 2]. Natural language processing, in this specific case text mining, is a toolbox that helps to obtain information from texts of financial analysts that may not be expressed by the quantitative estimates issued by analysts. This paper aims to explain how textual information in analyst reports provides market participants with an instrument to support their trading decisions. Ultimately, we intend to emphasize how previously rarely or not at all used textual information might potentially increase market efficiency.

Until now, it has not yet been examined to what extent textual information in analyst reports is informative about long-term investment returns. Therefore, different to event study settings in the previous literature [1–3] and different to portfolio strategies relying

Timo Schäfer thanks the efl – the Data Science Institute for financial support.

B. Clapham and J.-A. Koch (Eds.): FinanceCom 2020, LNBIP 401, pp. 59–74, 2020.
https://doi.org/10.1007/978-3-030-64466-6_4

on quantitative analyst opinion [4–6], our paper takes an investor's perspective at the calendar-year level to examine the profitability of portfolio strategies relying on the qualitative content in analyst reports.

We utilize the sentiment conveyed in analyst reports covering companies of the Dow Jones Industrial Average Index (DJIA) to construct quintile portfolios with different levels of sentiment for the time frame 2010 to 2018. Then, we investigate whether portfolios with varying levels of sentiment exhibit differences in performance and whether the outperformance is significant by taking several factor models into account. Also, we distinguish between brokerage firms with a strong and a weak track record in terms of price target forecasting error and additionally change our analysis setting from a predictive to a contemporaneous perspective.

Our findings suggest that analyst reports contain relevant information to explain returns contemporaneously and (to a somewhat lesser extent) future returns. More specifically, our results provide evidence that portfolio returns are increasing in the portfolio's average sentiment suggesting that the qualitative, textual content conveyed in analyst reports is reasonably informative. When considering a four-factor model [7] for the computation of portfolio returns, a portfolio that is long on stocks in the most positive tone quintile and short in stocks in the most negative tone quintile generates an average monthly return of 0.7%, which is significant at the 5% level. Our findings are robust to differences in analyst coverage and timely updates of analyst reports across sentiment portfolios but are highly correlated with a portfolio's average buy/hold/sell recommendation issued by analysts following. When differentiating between brokers' forecast errors with respect to price targets, we find that portfolio strategies using only reports from brokerage firms with relatively low average errors are more profitable than reports from brokers with high price target errors.

We contribute to the academic literature in several ways. First, we add to the literature that analyzes the value of analyst estimates and its implications for profitable investments [4, 5, 8–12]. When relying on the informational value of analyst consensus recommendations for the creation of profitable portfolio strategies, investors may realize excess returns due to high-frequent rebalancing of their portfolios that diminish when taking transaction costs into account [8]. Also, it has empirically been shown that more accurate analysts in terms of their earnings per share forecast accuracy also provide, on average, more profitable stock recommendations [6, 9].

Second, we contribute to the research stream that examines the informational value of text contained in analyst reports. Analysts evaluate company and industry-specific topics and comment on their earnings estimates, price targets, and recommendations. It has been demonstrated that financial analysts act as information intermediaries and help to better understand corporate information [2, 3, 12]. Furthermore, qualitative content contained in analyst reports provides information that goes beyond what is expressed by earnings estimates, price targets, and recommendations [13], and might be leveraged to measure corporate innovation [14].

Third, our paper is located in the finance sub-domain that analyzes stock return predictability exploiting text. For example, the volume on internet stock message boards predicts future stock returns [15]. Also, research that examines the relationship between news media sentiment and stock markets arrives at the conclusion that negative media sentiment can predict declining market prices [16]. Beyond that, variation in stock returns can successfully be predicted by text data from news [17].

This paper is structured as follows. Section 2 describes the data collection and preparation. Section 3 introduces the methodology for how we turn raw text data from analyst reports into portfolio strategies and evaluate their performance. Section 4 presents the results and Sect. 5 discusses our findings and concludes the paper.

2 Data

Our data sample comprises constituents of the DJIA from January 2010 to December 2018. Since the composition of the DJIA changes over time, our data set contains 36 different companies. We obtain analyst reports from Thomson One. We download analyst estimates, i.e., price targets and recommendation data, from the Institutional Brokers Estimate System (IBES) via the Wharton Research Data Services (WRDS) database. Specifically, we download the 12-month price targets to later calculate the forecast error per broker. Here, we match analyst reports and IBES price targets by using the respective company's ticker symbol to daily stock market data that comes from the Center for Research in Security Prices (CRSP).

We download 88,609 company-related research documents and filter for continuous text. We transform the text to lower case and tokenize it. Also, we remove numbers, stop words, and words shorter than three characters. In the following, we describe how we further process the data set to arrive at a stage where we exclusively deal with "true" analyst reports that can be matched with IBES estimates. In our view, "true" analyst reports are prepared manually by analysts and usually contain a text with a subjective assessment of a company, a recommendation, a price target, and earnings estimates. The sample creation process is as follows (see Table 1):

Table 1. **Sample creation process.** The steps are highlighted in bold in the text.

Step	Reason for the decrease in sample size	Resulting analyst reports
	Available research documents	*88,609*
1	Filter by metadata	67,826
2	Remove (near) duplicates	58,597
3	Remove short or corrupt analyst reports	51,286
4	Filter for matching ID with IBES	50,964
5	Remove inappropriate brokerage firms	49,243
6	Filter for reports of current DJIA constituents	42,137
7	Filter for price target per broker and company	31,524

- **Step 1**: We use metadata, e.g., company name, brokerage firm name, author name, and title to filter out those documents that obviously cannot be analyst reports.
- **Step 2**: We filter the data set for duplicates and near-duplicates by removing reports with 80% of the paragraphs matching those of a previous report. Thus, a report must have been changed by a minimum part to be used. Since there is no guideline for the threshold, we set a value ourselves after a thorough manual check.
- **Step 3**: We delete documents for which errors occurred during the conversion of the PDF files, e.g., text artifacts and missing spaces, and we filter reports for which no English language is detected and which are shorter than 250 words.
- **Step 4**: We remove analyst reports of brokers that we cannot match to IBES, e.g., because they do not report to IBES.
- **Step 5**: We identify brokerage firms whose reports do not meet a minimum quality standard (see definition of "true" analyst reports in the previous paragraph). Also, we filter out brokers that publish less than 10 reports during the observation period.
- **Step 6**: Since our trading strategy considers for each trading day those companies that are currently part of the DJIA, we filter the corresponding reports accordingly. For companies newly included in the DJIA, we keep the last reports published by each broker within a maximum period of 120 days prior to inclusion in the DJIA.
- **Step 7**: We match the analyst reports with the available IBES price targets per company. Since our analytical approach considers the price target performance error of the last two years, we filter for price targets that were published in the period three years to at least one year before the respective analyst report.

Table 6 (see Appendix) addresses the publication frequency of analyst reports and reveals that the average number of days between the publication of reports (*report distance*) is 3.69 days across all companies. This number is naturally also driven by the fact that many reports are regularly published in certain weeks. However, the variable *maximum report distance* helps to better understand the *report distance*, as it clarifies that only in rare cases no new reports between earnings results (the difference is usually 90 days) are published. From Table 6 it can also be seen that the analyst reports with an average of 1,213 words offer sufficient text for meaningful sentiment analysis. Furthermore, we suppose that the quality of the information contained in the analyst reports can also be positively attributed to the fact that, on average, 26.64 brokerage firms have published reports on the companies in our analysis (see Table 6). Additionally, the top 40 brokerage firms, based on the number of analyst reports in the data set, published 30,784 analyst reports, compared to overall 31,524 analyst reports.

3 Methodology

This section describes how we extract information from the text in analyst reports, construct quintile portfolios based on the information extracted, and how we evaluate the performance of each of the five portfolio strategies. The basic idea of our methodology is to construct daily mean sentiment scores for every constituent of the DJIA using the textual content of company-specific analyst reports. Then, we group companies into quintiles based on their daily mean sentiment and rebalance the quintile portfolios on a daily frequency when there is a change in a company's mean sentiment. Finally, we investigate whether the analyst sentiment is informative about companies' future and contemporaneous profitability.

3.1 Trading Signals

The main idea is to extract textual information from analyst reports and transform the preprocessed text into a numerical representation, e.g., a single value. We construct trading signals from the texts of analyst reports that previous research has identified to be informative about the (future) performance of the respective company. Specifically, accounting and finance research utilizes sentiment scores computed by word lists [16, 18]. The usage of these scores allows us to make a statement on whether document A is more positive or optimistic than document B. We define the sentiment score of analyst a regarding company i at day t as

$$sentiment_{i,a,t} = \frac{\#positive_{i,a,t} - \#negative_{i,a,t}}{\#positive_{i,a,t} + \#negative_{i,a,t}}, \tag{1}$$

where $\#positive_{i,a,t}$ and $\#negative_{i,a,t}$ count the number of positive and negative words in the respective analyst's report using the word lists introduced in [18]. Other analyst-related studies determine the sentiment at the sentence level for analyst reports using a naïve Bayes approach [2, 3] or might be domain-specific language representation models trained for sentiment classification. However, we prefer the flexibility of a dictionary approach in our case, which does not require pre-trained data.

3.2 Portfolio Construction

Equation 1 allows us to calculate the daily sentiment of a company i as indicated by a specific analyst a. Since companies are usually covered by one analyst per brokerage firm, this also corresponds to the brokerage firm opinion. To get a company-day-level representation, we aggregate these sentiment scores by computing the daily average sentiment of a company i by taking all analysts into account that follow company i. In particular, for every trading day t, we compute for each company i the average sentiment score of the set of the most current reports issued by analysts that follow company i, i.e., the last available analyst report for every brokerage firm:

$$\overline{sentiment_{i,t}} = \frac{1}{n_{i,t}} \sum_{a=1}^{n_{i,t}} sentiment_{i,a,t}, \tag{2}$$

where $n_{i,t}$ is the number of analysts following company i at day t. We discard analyst reports at day t that are older than 120 calendar days assuming that the information in these reports does not represent the current information environment [4]. At each trading day t, we include the current set of constituents of the DJIA.

We construct five portfolios by sorting companies into quintiles based on their average sentiment defined in Eq. 2 [8, 9]. For instance, the first portfolio contains the six companies with the highest average sentiment scores on day t and the fifth portfolio contains the six companies with the lowest average sentiment score on day t. For each of the five portfolios p, we calculate daily value-weighted returns $R_{p,t}$ using stock i's market capitalization at day $t - 1$:

$$R_{p,t} = \sum_i mcap_{i,t-1}R_{i,t}, \quad p = 1, \ldots, 5, \tag{3}$$

where $R_{i,t}$ is the daily return of stock i at day t and $mcap_{i,t-1}$ is the market capitalization of stock i at $t - 1$ divided by the aggregate market capitalization of all firms in portfolio p at $t - 1$. To compute portfolio returns as in Eq. 3, we use the composition of portfolio p at day $t - 1$ to compute the return at day t, $R_{p,t}$ [8]. This allows us to have a predictive perspective in our quintile portfolios. We start our analysis in March 2010 such that we can compute mean sentiment scores via Eq. 3 using analyst reports from January and February 2010. Otherwise, our mean sentiment scores in January and February would not have sufficient coverage of analyst reports.

We use the daily returns for each strategy p to create monthly returns $R_{p,m}$:

$$R_{p,m} = \left(\prod_{t=1}^{m_t} \left(1 + R_{p,t} \right) \right) - 1, \tag{4}$$

where m_t is the number of trading days in the respective month m. Supposing that analyst reports are, on average, informative, we expect that the text-based sentiment is informative about the future performance of the respective companies and can be exploited in terms of a trading strategy.

3.3 Impact of Forecast Error

Analysts with a lower error regarding earnings per share (EPS) forecasts provide more profitable recommendations, which holds when controlling for a set of other determinants of investment profitability, e.g., analyst expertise [5, 9]. Building on that, we scrutinize whether analysts who show lower forecasting errors (historically) publish reports that are more informative about future investment profitability. We measure analysts' error rates using price targets they issue as price targets have a clear forecast horizon of (mostly) 12 months and their forecast error can be easily computed. The absolute target price error (ATPE) [19] is defined as

$$ATPE_{i,a,t} = \frac{\left| TP_{i,a,t} - P_{i,t+12} \right|}{P_{i,t}}, \tag{5}$$

where $TP_{i,a,t}$ is the target price of analyst a at issue day t regarding company i, $P_{i,t+12}$ is the price of company i 12 months after the release of the target price $TP_{i,a,t}$, and $P_{i,t}$ is the price of company i at day t serving as a scaling factor. We determine $P_{i,t+12}$ by going 250 business days ahead from the issuance day t.

For each stock i, we compute two mean sentiment scores using Eq. 2 – one comprises brokers with high absolute forecast errors, i.e., above the median ATPE, and one with low errors, i.e., equal to or below the median ATPE. For each calendar year y, we compute a broker's ATPE across all companies in our DJIA sample based on the mean ATPE in years $y - 3$ and $y - 2$. As we take a predictive perspective, we do not want to incorporate price information from calendar year y or any succeeding year. If we include ATPE information from year $y - 1$, we consider price information from year y and bias our analysis. This procedure relies on findings that document a persistence in analysts' price target performance over time [6].

3.4 Performance Evaluation

We examine whether trading on sentiment signals from analyst reports generates positive and significant returns. In particular, we compare the performance of each quintile strategy and whether quintiles with more distinct signals, i.e., top and bottom quintiles, generate more significant returns of higher absolute magnitude. We construct portfolios using long and long-short strategies. The long-short strategy is a zero-investment strategy that represents the difference between the daily returns of a long portfolio and the daily returns of a short portfolio.

We compute a strategy's raw return and market-adjusted return first. Also, we calculate market-adjusted returns by subtracting the daily value-weighted return on all NYSE, AMEX, and NASDAQ stocks from the raw returns. Then, we compute a strategy's alpha using the one-factor Capital Asset Pricing Model (CAPM), the three-factor Fama-French model [20], as well as the four-factor model [7] to account for several stock market factors.[1] We estimate the CAPM regression as follows:

$$R_{p,t} - RF_t = \alpha_p + \beta_{1,p}RMRF_t + \epsilon_{p,t}, \tag{6}$$

the three-factor Fama-French model [20] as

$$R_{p,t} - RF_t = \alpha_p + \beta_{1,p}RMRF_t + \beta_{2,p}HML_t + \beta_{3,p}SMB_t + \epsilon_{p,t}, \tag{7}$$

and the four-factor model [7] as

$$R_{p,t} - RF_t = \alpha_p + \beta_{1,p}RMRF_t + \beta_{2,p}HML_t + \beta_{3,p}SMB_t + \beta_{4,p}UMD_t + \epsilon_{p,t}, \tag{8}$$

where RF_t is the risk-free rate (one-month treasury bill rate), $RMRF_t$ is the value-weighted return on all NYSE, AMEX, and NASDAQ stocks minus the risk-free rate, HML_t is the average return on two value (high book-to-market ratio) portfolios minus the average return on two growth (low book-to-market ratio) portfolios, SMB_t is the average return on three small (in terms of company size) portfolios minus the average return on three big (in terms of company size) portfolios, and UMD_t is the average return on a momentum portfolio at day t. The intercept α_p is the coefficient of interest, that is, the (monthly) alpha generated by the respective portfolio strategy $p(p = 1, \ldots, 5)$.

4 Results

The previous section has introduced the methodology of how we construct quintile portfolios based on companies' daily mean sentiment scores from analyst reports. Table 2 provides descriptive statistics on the distribution of daily mean sentiment scores of companies that are in each of the five portfolios as defined in Eq. 2 at the company-day level. Panel A exhibits all analyst reports while Panels B and C refer to analyst reports of brokerage firms with a high and a low ATPE (as defined in Eq. 5), respectively. Strategy 1 refers to the portfolio that contains the companies with the lowest sentiment scores (bottom quintile), while strategy 5 represents the portfolio with companies that exhibit the highest sentiment scores (top quintile).

---

[1] We thank Kenneth French for providing data on stock market factors, available at http://mba. tuck.dartmouth.edu/pages/faculty/ken.french/data_library.html.

In Table 2, we see that portfolio 5 in Panel A with the most favorable tone in terms of sentiment exhibits – by construction – the highest mean sentiment score, which is nearly twice as high as in the second most favorable portfolio 4. (Nearly) none of the company-day observations in these two portfolios have a non-positive score, as their value for the 1% percentile is 0.05 and 0.16, respectively. Portfolio 2 is relatively neutral as its average sentiment score is around 0, which means that the fraction of positive and negative words in the analyst reports is, on average, largely balanced. In contrast, the mean sentiment score of portfolio 1, which contains the least favorable companies in terms of sentiment, is very negative and, most importantly, its value for the 75%-percentile is highly negative as well, which implies that a large fraction of the company-day observations in this portfolio exhibits a negative sentiment score. The interpretations of the descriptive statistics on Panel A also qualitatively hold for Panels B and C. In the following, we are primarily interested in the two strategies with the most distinct information signals, i.e., strategies 1 and 5.

Table 2. Summary statistics on mean sentiment scores across quintile portfolios. The column count denotes the number of day-stock observations for each strategy. The %-columns refer to the percentiles. Panel A includes the sentiment of all analyst reports, Panel B (Panel C) includes analyst reports of brokers with ATPEs (see Eq. 5) larger (equal to or smaller) than the median for a given company in a specific calendar year.

Strategy	Count	Mean	Std.	1%	25%	50%	75%	99%
Panel A: All
1	13,562	−0.21	0.12	−0.56	−0.28	−0.2	−0.13	0.02
2	13,546	−0.03	0.07	−0.2	−0.08	−0.03	0.02	0.14
3	13,210	0.08	0.06	−0.06	0.04	0.08	0.13	0.23
4	13,546	0.19	0.07	0.05	0.15	0.19	0.24	0.38
5	13,555	0.37	0.11	0.16	0.29	0.36	0.45	0.64
Panel B: High error
1	13,477	−0.25	0.13	−0.58	−0.33	−0.23	−0.15	0.0
2	13,463	−0.04	0.08	−0.23	−0.09	−0.03	0.02	0.17
3	13,069	0.09	0.07	−0.09	0.04	0.08	0.13	0.28
4	13,462	0.21	0.08	0.05	0.15	0.21	0.27	0.4
5	13,466	0.41	0.11	0.18	0.32	0.4	0.48	0.69
Panel C: Low error
1	13,477	−0.26	0.14	−0.67	−0.33	−0.24	−0.15	−0.01
2	13,443	−0.05	0.08	−0.25	−0.1	−0.05	0.0	0.13
3	13,030	0.08	0.07	−0.08	0.03	0.08	0.13	0.25
4	13,443	0.21	0.08	0.03	0.15	0.2	0.27	0.39
5	13,470	0.4	0.12	0.16	0.31	0.39	0.48	0.69

Now, we examine the performance in terms of returns of the portfolio strategies. For each strategy, we compute daily returns that we transform into monthly returns using Eq. 4. This perspective is predictive as we only consider information until yesterday $(t-1)$ to calculate the returns for today (t). We estimate Eq. 8 to examine which factors drive the return variation in each of the quintile strategies for the Panels (see Table 3).

Table 3. Results from four-factor model estimations using monthly data. All portfolios follow a long strategy, except for 5–1 that is long in strategy 5 and short in strategy 1. t-values are reported in parentheses and are corrected for heteroscedasticity and autocorrelation. *, **, and *** represent statistical significance at the 10%, 5%, and 1% level. The levels of significance represent whether the estimated coefficients are different from zero (1 for RMRF).

Strategy	RMRF	SMB	HML	UMD	Adj. R^2
Panel A: All					
1	1.01 (0.08)	−0.2 (−1.95)*	0.26 (2.39)**	−0.17 (−2.16)**	0.75
2	0.93 (−1.25)	−0.4 (−4.68)***	0.18 (1.95)*	−0.05 (−0.69)	0.75
3	0.81 (−2.89)***	−0.17 (−1.61)	0.06 (0.53)	−0.07 (−0.82)	0.62
4	0.87 (−1.99)**	−0.47 (−4.5)***	−0.01 (−0.12)	0 (−0.01)	0.62
5	0.91 (−1.73)*	−0.24 (−2.83)***	−0.01 (−0.1)	0.2 (2.97)***	0.75
5–1	−0.09 (−12.9)***	−0.04 (−0.29)	−0.26 (−1.83)*	0.37 (3.44)***	0.18
Panel B: High error					
1	0.99 (−0.14)	−0.22 (−2.17)**	0.17 (1.53)	−0.13 (−1.61)	0.72
2	0.84 (−2.95)***	−0.22 (−2.55)**	0.28 (3.07)***	−0.1 (−1.4)	0.73
3	0.89 (−1.68)*	−0.42 (−3.84)***	0.12 (1.02)	0.05 (0.57)	0.62
4	0.93 (−1.24)	−0.47 (−5.42)***	−0.06 (−0.64)	0.04 (0.57)	0.73
5	0.87 (−2.44)**	−0.19 (−2.24)**	−0.02 (−0.25)	0.07 (1.11)	0.74
5–1	−0.12 (−12.7)***	0.04 (0.27)	−0.19 (−1.26)	0.21 (1.84)*	0.06
Panel C: Low error					
1	1.04 (0.57)	−0.22 (−2.05)**	0.25 (2.18)**	−0.07 (−0.75)	0.72
2	0.83 (−2.73)***	−0.27 (−2.77)***	0.16 (1.55)	−0.14 (−1.82)*	0.66
3	1.05 (0.79)	−0.49 (−5.18)***	0.08 (0.76)	−0.03 (−0.41)	0.75
4	0.82 (−3.78)***	−0.31 (−3.99)***	−0.03 (−0.31)	0.04 (0.6)	0.73
5	0.85 (−2.71)***	−0.23 (−2.68)***	0.01 (0.05)	0.15 (2.16)**	0.7
5–1	−0.19 (−13.25)***	−0.01 (−0.07)	−0.25 (−1.63)	0.22 (1.9)*	0.11

Regarding the market betas (coefficients of RMRF), there is mixed evidence as some strategies exhibit a factor loading that is statistically not distinguishable from 1, and the majority of strategies have a loading distinguishable from 1 (at the 10% level). This implies that most of the strategies largely do not correspond to movements in the stock market in general. Nearly all coefficients for the SMB factor are significantly different from zero and negative for all strategies. Thus, the set of strategies outlined here is indicative of large stocks, which is reasonable for our DJIA sample.

Next, we scrutinize the long-term profitability of textual information in analyst reports. Table 4 shows the results of the quintile strategies using monthly returns. Considering all reports of a company (Panel A), our results on the raw return and market-adjusted return (first two columns) show that the average sentiment score of a portfolio is predictive for its profitability. There is a higher raw and market-adjusted return with an increasing, relatively more positive sentiment from strategy 1 to 5 (except for strategy 2).

Table 4. Performance across quintile portfolios – predictive. Raw returns are average monthly returns as defined in Eq. 4. Market-adjusted returns are computed as raw returns minus the monthly value-weighted return on all NYSE, AMEX, and NASDAQ stocks. CAPM, three-factor, and four-factor models are the estimated monthly alphas (intercepts) using Eqs. 6, 7, and 8, respectively. All portfolios follow a long strategy, except for 5–1 that is long in strategy 5 and short in strategy 1. Panel A includes the sentiment of all analyst reports, Panel B (Panel C) includes analyst reports of brokers with ATPEs (see Eq. 5) larger (equal to or smaller) than the median for a given company in a specific calendar year. t-values are reported in parentheses and are corrected for heteroscedasticity and autocorrelation. *, **, and *** represent statistical significance at the 10%, 5%, and 1% level.

strategy	raw return	market-adj. return	CAPM	three-factor	four-factor
Panel A: All					
1	0.182	−0.783	−0.67	−0.659	−0.59
	(0.427)	(−3.409)***	(−2.395)**	(−2.486)**	(−2.248)**
2	0.797	−0.169	−0.258	−0.305	−0.287
	(2.157)**	(−0.801)	(−1.056)	(−1.386)	(−1.291)
3	0.648	−0.317	−0.194	−0.213	−0.186
	(1.806)*	(−1.343)	(−0.717)	(−0.79)	(−0.685)
4	0.784	−0.181	−0.232	−0.326	−0.326
	(2.148)**	(−0.698)	(−0.796)	(−1.209)	(−1.195)
5	1.335	0.37	0.245	0.187	0.108
	(3.753)***	(1.844)*	(1.082)	(0.847)	(0.505)
5–1	1.153	1.153	0.914	0.846	0.698
	(3.606)***	(3.606)***	(2.43)**	(2.32)**	(2.002)**
Panel B: High Error					
1	0.32	−0.645	−0.397	−0.41	−0.356
	(0.771)	(−2.846)***	(−1.452)	(−1.547)	(−1.348)
2	0.619	−0.346	−0.362	−0.354	−0.316
	(1.737)*	(−1.62)	(−1.483)	(−1.573)	(−1.4)
3	0.709	−0.257	−0.511	−0.574	−0.595
	(1.883)*	(−0.992)	(−1.775)*	(−2.114)**	(−2.169)**
4	0.862	−0.104	−0.068	−0.172	−0.188
	(2.389)**	(−0.469)	(−0.265)	(−0.761)	(−0.822)
5	0.997	0.032	−0.194	−0.235	−0.266
	(2.875)***	(0.163)	(−0.934)	(−1.143)	(−1.283)
5–1	0.677	0.677	0.202	0.174	0.091
	(2.181)**	(2.181)**	(0.571)	(0.494)	(0.259)
Panel C: Low Error					
1	0.25	−0.715	−0.837	−0.836	−0.81
	(0.572)	(−2.953)***	(−2.839)***	(−2.946)***	(−2.827)***
2	0.564	−0.401	−0.515	−0.532	−0.476
	(1.555)	(−1.687)*	(−1.925)*	(−2.091)**	(−1.878)*
3	0.964	−0.002	−0.172	−0.254	−0.242
	(2.364)**	(−0.007)	(−0.631)	(−1.038)	(−0.978)
4	0.747	−0.218	−0.019	−0.087	−0.102
	(2.343)**	(−1.085)	(−0.091)	(−0.432)	(−0.5)
5	1.095	0.129	0.164	0.111	0.05
	(3.197)***	(0.613)	(0.697)	(0.478)	(0.22)
5–1	0.845	0.845	1.001	0.947	0.861
	(2.606)***	(2.606)***	(2.618)***	(2.514)**	(2.297)**

Raw returns are statistically significant for strategies 2 to 5, market-adjusted returns for strategies 1 and 5. The monthly average market-adjusted return from the zero-investment strategy 5–1 amounts to 1.153% and is statistically significant at the 1% level. Here, the raw return is equivalent to the market-adjusted return as the market return cancels out. However, raw returns are either negative or not significant anymore in strategies 2 to 4, when adjusting for market movements.

For the factor models, i.e., CAPM, three-factor Fama-French model, and four-factor model, Table 4 contains the respective results in columns CAPM, three-factor, and four-factor. When considering these further stock market factors, our results indicate that the estimated intercept of the factor models – the monthly alpha – is zero in nearly all specifications and becomes nearly statistically insignificant. An exception is strategy 1 comprising stocks with the lowest sentiment, which generates a significant negative alpha in all return model specifications. Strategy 5 is the only strategy that still generates a positive (but insignificant) alpha. Our results suggest that the textual content in analyst reports is predictive for future stock returns since the zero-investment strategy 5–1 generates a monthly alpha of 0.698% that is significant at the 5% level. This implies that investors can to some extent exploit information in analyst reports for long-term investment returns.

For Panels B and C, and in line with the literature [5, 9], we find that brokers with a low ATPE convey more profitable information in their reports than brokers with a high ATPE. In particular, when comparing the returns for the strategy 5–1 in Panel B with Panel C, we see that for all model specifications the returns in Panel C are higher than in Panel B. Most importantly, information from brokers with low price target errors in Panel C generates a significant monthly alpha of 0.86% as compared to about (insignificant) 0.1% in the case of Panel B.

Table 5. Summary statistics on rebalancing magnitude across quintile portfolios. The table shows statistics on the daily mean number of changes in each strategy's portfolio using reports of all brokers. For instance, when in strategy p stock i drops out and stock i' comes in from day $t - 1$ to day t, then the rebalancing score is 1 at day t for strategy p. The column count denotes the number of company-day observations for each strategy. The %-columns refer to the percentiles.

Strategy	Count	Mean	Std.	1%	25%	50%	75%	99%
1	2,263	0.33	0.52	0.0	0.0	0.0	1.0	2.0
2	2,263	0.7	0.74	0.0	0.0	1.0	1.0	3.0
3	2,263	0.8	0.78	0.0	0.0	1.0	1.0	3.0
4	2,263	0.74	0.76	0.0	0.0	1.0	1.0	3.0
5	2,263	0.34	0.51	0.0	0.0	0.0	1.0	2.0

Our findings illustrate that the qualitative information in analyst reports is informative for future stock returns, as we can order portfolio returns by magnitude and there is a significant portfolio return after adjusting for market movements. Our results suggest that investors can exploit negative information, i.e., portfolio strategy 1 with the least positive companies, more effectively regarding investment opportunities. This is in line with Huang et al. [2] who conclude that investors weigh bad news more than twice as strong as positive news in analyst reports. In contrast to Loh and Mian [9], the profits that our quintile portfolio strategy generates can be realized by investors as this strategy relies on all information available at trading day t (and on no future information at day $t' > t$) when constructing the portfolio and, thus, is predictive on the future stock price development.

To also understand how well the information in analyst reports can explain contemporaneous returns, we slightly alter the definition of Eq. 3 and use the information available at day t instead of $t - 1$ to construct portfolios. The results reveal that returns (in Panel A) in all five specifications can be again ordered in a (nearly) monotonous way giving confidence that the text in analyst reports describes stock returns contemporaneously. In contrast to the predictive approach in Table 4, the two positive (regarding sentiment) strategies 4 and 5 exhibit positive alphas, though not statistically significant in all specifications. However, monthly alphas generated by strategy 5 are significant in the case of a four-factor model at the 10% level. Importantly, the alpha of the long-short strategy 5–1 is highly significant at the 1% level and accumulates to about 1.66% per month. Also, the results qualitatively hold for Panels B and C compared to the predictive setting.

Thus, our empirical findings suggest that analysts react, on average, to the revelation of new information and incorporate this information in their reports and to a lesser extend provide predictive and forward-looking information. Since our analysis does not take trading costs into account, we will at least consider how often rebalancing has been necessary. Table 5 describes the magnitude of daily rebalancing for each of the quintile strategies. We see that rebalancing is, on average, much less frequent (median is zero) and of lower magnitude (mean is less than half as large as for strategies 2 to 4) for the top and bottom quintile strategies that mainly drive our results. This lends support to our argumentation that the information in the top and bottom quintiles is more persistent as we observe fewer changes in the portfolio composition.

5   Discussion and Conclusion

This paper demonstrates that information from textual parts of analyst reports can be used profitably in the long run. For the implementation of our findings, we see two possible fields of application in practice. On the one hand, our findings can be integrated into trading strategies and thus create added value for the investment industry. On the other hand, we see potential in the implementation of our findings for company risk management and insurance companies. From a theoretical perspective, we provide evidence that it is not necessary to focus exclusively on quarterly earnings results, which is the case in other analyst-related studies. By showing the potential of analyst texts for a predictive trading setting, we also highlight the fact that research needs to address the texts of specific disciplines more closely. Moreover, we see a particularly important contribution to research in the fact that we can confirm that it is possible to separate analysts based on their forecast accuracy and thus gain an investment advantage. Until now, this has only been done based on quantitative measures [5, 6, 9]. In our case, we show that this is also reflected in the profitability of the text-based trading strategies of the groups classified by the ATPE.

Other studies, which in contrast to our study deal with non-textual analyst data [8, 9], rely on larger samples. Against this background, our results potentially suffer from selection bias, as we only consider companies that are constituents of the DJIA, i.e., we cannot generalize our findings to other stocks that have a lower market capitalization. To prevent another potential selection bias, we also run our analysis without filtering for brokers included in IBES. However, our results remain qualitatively unchanged. Moreover, our findings do not account for trading costs. The consideration of trading costs likely diminishes the reported alpha as our portfolio strategy relies on daily rebalancing [8]. Our findings show that our strategy is not strongly affected by too frequent rebalancing. Our approximation of the information revealed in analyst reports using dictionary-based methods is a widely used, straightforward method, but machine-learning sentiment classifications may lead to improved results [2]. As to whether this approach proves beneficial, future research needs to address. Besides, our results are robust to an alternative formulation of our sentiment formula, which uses the total number of words instead of the total number of polarity words in the denominator. Additionally, one might argue that we use analyst reports from a specific day without considering the time of release, e.g., after the closing of stock exchanges. We address this by using the sentiment scores of $t - 2$ instead of $t - 1$ to construct portfolios. We find that the results are robust to a delay in investment.

Overall, this paper sheds light on the investment value of qualitative information in analyst reports. To the best of our knowledge, this is the first research project that uses textual analyst data for day-level long-term investment strategies. Specifically, our paper provides evidence that portfolio returns are increasing in the average sentiment suggesting that qualitative content communicated in analyst reports is informative. When accounting for a four-factor model [7], a predictive zero-investment strategy that is long on the stocks with the most positive analyst reports and short on the stocks with the least positive reports generates a significant monthly alpha of about 0.7%. Our results imply that analysts provide information in their reports that explain contemporaneous returns but most importantly have predictive power. This finding might be particularly useful since analyst reports usually provide well-researched information which, in contrast to their less frequently adjusted quantitative predictions, are published at relatively short intervals (see Table 6 in the Appendix). Therefore, future research should scrutinize to what extent qualitative information is incremental to quantitative recommendations or price target data regarding investment profitability. Apart from this, future research may validate our results based on a larger data set, which allows further generalization, and by comparing further approaches to calculate sentiment scores.

Appendix

Table 6. Summary statistics on the analyst report sample. The *report distance* indicates the average number of days between analyst reports. The column *brokers* lists how many different brokers have published reports per company.

Company	Reports	Report distance	Max. report distance	Avg. words per report	Brokers
3 M	595	5.50	82	1,453.89	20
AT&T	785	2.41	42	1,231.49	28
Alcoa	207	6.53	73	1,016.13	13
American Express	1,102	2.97	45	939.27	27
Apple	1,186	1.20	18	1,186.05	31
Bank of America	413	3.26	39	1,559.00	17
Boeing	1,342	2.44	27	994.23	29
Caterpillar	1,235	2.64	39	918.11	23
Chevron	821	3.98	53	1,371.20	24
Cisco Systems	1,605	2.04	40	1,229.03	41
Coca-Cola	734	4.42	77	1,166.94	25
E I du Pont	536	5.16	75	1,036.19	19
Exxon	694	4.67	75	1,405.35	25
General Electric	1,021	3.02	48	1,673.06	18
Goldman Sachs	530	3.77	63	1,076.86	23
Hewlett-Packard	382	3.54	53	1,587.48	16
Home Depot	994	3.27	79	1,068.87	30
Intel	1,987	1.65	55	1,104.18	50
IBM	876	3.71	89	1,103.40	33
JPMorgan Chase	1,082	3.02	57	1,520.37	29
Johnson & Johnson	1,143	2.85	69	1,262.01	27
Kraft Foods	75	13.07	86	1,803.84	11
McDonald's	1,167	2.77	66	984.52	32

(*continued*)

Table 6. (*continued*)

Company	Reports	Report distance	Max. report distance	Avg. words per report	Brokers
Merck & Co	1,051	3.10	63	983.33	26
Microsoft	1,421	2.31	72	1,283.89	37
Nike	658	3.05	84	1,197.77	27
Pfizer	951	3.44	50	1,010.11	25
Procter & Gamble	858	3.80	63	1,337.28	30
Travelers Companies	530	6.13	90	1,064.55	23
United Technologies	838	3.89	56	1,365.89	23
UnitedHealth	854	2.72	43	1,306.38	27
Verizon	1,031	3.17	74	1,077.40	32
Visa	616	3.29	83	942.90	31
Walgreens Boots Alliance	71	3.83	55	1,029.39	15
Walmart	1,285	2.55	58	1,246.57	36
Walt Disney	848	3.82	66	1,137.71	36
Mean	875.67	3.69	61.31	1,213.18	26.64
Sum	31,524				

References

1. Twedt, B., Rees, L.: Reading between the lines: an empirical examination of qualitative attributes of financial analysts' reports. J. Account. Public Policy **31**(1), 1–21 (2012)
2. Huang, A.H., Zang, A.Y., Zheng, R.: Evidence on the information content of text in analyst reports. Acc. Rev. **89**(6), 2151–2180 (2014)
3. Huang, A.H., Lehavy, R., Zang, A.Y., Zheng, R.: Analyst information discovery and interpretation roles: a topic modeling approach. Manage. Sci. **64**(6), 1–23 (2017)
4. Womack, K.L.: Do brokerage analysts' recommendations have investment value? J. Financ. **51**(1), 137–167 (1996)
5. Ertimur, Y., Sunder, J., Sunder, S.V.: Measure for measure: the relation between forecast accuracy and recommendation profitability of analysts. J. Acc. Res. **45**(3), 567–606 (2007)
6. Bradshaw, M.T., Brown, L.D., Huang, K.: Do sell-side analysts exhibit differential target price forecasting ability? Rev. Acc. Stud. **18**(4), 930–955 (2013)
7. Carhart, M.M.: On persistence in mutual fund performance. J. Financ. **52**(1), 57–82 (1997)
8. Barber, B., Lehavy, R., McNichols, M., Trueman, B.: Can investors profit from the prophets? Security analyst recommendations and stock returns. J. Financ. **56**(2), 531–563 (2001)
9. Loh, R.K., Mian, G.M.: Do accurate earnings forecasts facilitate superior investment recommendations? J. Financ. Econ. **80**(2), 455–483 (2006)
10. Bradshaw, M.T.: How do analysts use their earnings forecasts in generating stock recommendations? Acc. Rev. **79**(1), 25–50 (2004)

11. Bradshaw, M.T.: Analysts' forecasts: what do we know after decades of work? Available at SSRN 1880339 (2011)
12. Bradley, D., Clarke, J., Lee, S., Ornthanalai, C.: Are analysts' recommendations informative? Intraday evidence on the impact of time stamp delays. J. Financ. **69**(2), 645–673 (2014)
13. Kerl, A.G., Walter, A.: Never judge a book by its cover: what security analysts have to say beyond recommendations. Fin. Markets. Portfolio Mgmt. **22**(4), 289–321 (2008)
14. Bellstam, G., Bhagat, S., Cookson, J.A.: A text-based analysis of corporate innovation. Manage. Sci. (2020, Forthcoming)
15. Antweiler, W., Frank, M.Z.: Is all that talk just noise? The information content of internet stock message boards. J. Financ. **59**(3), 1259–1294 (2004)
16. Tetlock, P.C.: Giving content to investor sentiment: the role of media in the stock market. J. Financ. **62**(3), 1139–1168 (2007)
17. Manela, A., Moreira, A.: News implied volatility and disaster concerns. J. Financ. Econ. **123**(1), 137–162 (2017)
18. Loughran, T., McDonald, B.: When is a liability not a liability? Textual analysis, dictionaries, and 10-Ks. J. Financ. **66**(1), 35–65 (2011)
19. Bilinski, P., Lyssimachou, D., Walker, M.: Target price accuracy: international evidence. Acc. Rev. **88**(3), 825–851 (2012)
20. Fama, E.F., French, K.R.: Common risk factors in the returns on stocks and bonds. J. Financ. Econ. **33**, 3–56 (1993)

Alternative Trading and Investment Offerings by FinTechs

Portfolio Rankings on Social Trading Platforms in Uncertain Times

Steffen Bankamp$^{(\boxtimes)}$ and Jan Muntermann

University of Goettingen, Goettingen, Germany
steffen.bankamp@uni-goettingen.de,
muntermann@wiwi.uni-goettingen.de

Abstract. On social trading platforms, anyone can create investment portfolios and share them with others. This leads to an abundance of portfolio strategies available to investors. However, this variety comes with the drawback of high search costs for investors looking for the most promising portfolios to invest in. Platform operators provide a ranking on these portfolios, thus making it easier for investors and reducing search costs. In our study, we analyze the value of this ranking regarding various aspects and especially against the background of its protective function in turbulent market phases. For this purpose, we look at the stock market crash in connection with the Covid-19 pandemic. Our results show (i) that investors on social trading platforms are making use of the ranking, (ii) that the ranking offers value for these investors, and (iii) protects them from extreme losses in downward periods. The ranking we analyzed rather identifies portfolios with a more successful market timing than those with an effective stock selection. We further show that changes in the ranking tend to contribute to procyclical behavior (iv). Finally, we provide preliminary evidence that indicates rankings' influence on the behavior of traders.

Keywords: Social trading · Rankings · FinTech · Platforms · Volatile market environment

1 Introduction

Social trading platforms allow private individuals to act as fund managers and construct portfolios for other users. As everybody can construct these portfolios, a large number of available portfolios with different strategies is created. However, at the same time, this abundance leads to a major decision problem for the users who want to invest in these portfolios.

In order to reduce the search costs for these investors, social trading platforms have introduced rankings. These can be used by investors to reduce their choice set. Previous research has found that investors make use of these rankings to a large extent and portfolios listed on top of these rankings receive higher capital inflows [1]. Thus, a substantial part of the analysis looking for suitable portfolios is no longer performed by the investors, but by the platform itself. Similar approaches can be observed on other

© Springer Nature Switzerland AG 2020
B. Clapham and J.-A. Koch (Eds.): FinanceCom 2020, LNBIP 401, pp. 77–91, 2020.
https://doi.org/10.1007/978-3-030-64466-6_5

FinTech platforms. Operators of peer-to-peer lending platforms for example are strongly engaged in the evaluation of loans, while investors on these platforms are following a rather passive investment approach [2]. Since many investors rely on such rankings, it is highly relevant to examine which benefits these rankings provide to them.

The first social trading platform OpenBook was launched by eToro[1] in 2010 [3]. As this happened after the financial crisis, social trading platforms so far were operating in a market environment in which a large portion of security prices was rising. Consequently, there is little empirical evidence on the performance of social trading platforms in turbulent market phases and, more specifically, to what extent the portfolio rankings help to limit or control the risk of investors. The stock market crash at the beginning of 2020 in connection with the Covid-19 pandemic provides an unprecedented opportunity to evaluate the ability of such rankings, especially under these market conditions when the savings of investors are most vulnerable. Therefore, we ask the following research question:

RQ1: *Does the portfolio ranking provided by social trading platforms offer economic value to investors?*

If we find that rankings offer an economic benefit to investors, the reasons for this should be analyzed subsequently. For this purpose, we ask the following research question:

RQ2: *How does the portfolio ranking provide economic value to investors on social trading platforms?*

The paper is structured as follows. First, we will discuss the fundamentals of social trading. Then, we analyze the literature dealing with rankings on online platforms in general and social trading platforms in particular. Given this theoretical background, we derive our hypotheses. Afterward, the utilized dataset is presented. Within the analysis section, we compare the top-ranked portfolios with a control group to synthesize the use of such rankings. The paper concludes with a discussion and a summary of the main results.

2   Theoretical Foundations

2.1   Social Trading

Social trading platforms are combining properties of social media networks [4] with a new type of classical delegated portfolio management [5]. A social trading platform is a two-sided platform. *Traders* construct portfolios and publish them via the social trading platform. The *investors* in turn use the platform to find suitable portfolios and invest in them via copy trades. A copy trade can either refer to a copy of a single transaction conducted by the trader or to a copy of the entire trading strategy revealed by the trader over time [1]. The two participant groups, *traders* and *investors* are not exclusive. Thus, investors can also publish own portfolios. On some platforms, a social trading portfolio

---

[1] https://www.etoro.com.

itself can be part of another social trading portfolio. Furthermore, one trader can also operate several portfolios in parallel.

To reduce the information asymmetry between investors and traders, social trading platforms provide a detailed track record on each portfolio. This transparency exceeds the transparency usually provided in traditional delegated portfolio management via mutual funds [5]. The social media features of social trading platforms include profile pages, comments, forums, and chats. A detailed representation of the generic design of social trading platforms can be found in Glaser and Risius [6].

Traders are remunerated for their activity through the social trading platform. The costs are directly or indirectly carried by the investors. According to Doering et al. [5], three remuneration models have been established. These include follower-based, profit-based with high water mark, and volume-based compensation schemes.

Regarding the instruments traded on the platform, the social trading platforms show considerable diversity. However, currency trading is very common [6]. An example of a platform focusing on currency trading is ZuluTrade[2]. The eToro platform has a broader base and is used to trade foreign currencies, cryptocurrencies, and equities.

2.2 Rankings on Online Platforms

Search Rankings play a major role within the platform economy, as they are an important tool to match platform users and facilitate transactions [7]. These rankings make it easier for platform users to find an adequate counterpart and therefore decrease search costs, which in turn increases economic welfare [8]. The platform generates economies of scale in data collection and analysis. If the ranking is provided by the platform, it only needs to be carried out once instead of by each user individually. This also puts the platform in an advantageous position, which could be exploited by the platform through biased rankings that may manipulate the behavior of the platform users [7]. Furthermore, rankings promote lock-in effects as the top-ranked entities are receiving even more attention, which ultimately strengthens their top-ranked position [7]. Comprehensive literature dealing with the influence of rankings on the users of e-commerce platforms can be found. For instance, Ghose et al. [9] showed that rankings have a causal effect on the customers' click and purchase decision. In contrast, Ursu [8] only found that the ranking position casually affects the customers' search (click-through-rate) but not the purchase decision when controlling for the clicked items (conversion rate).

In addition to this literature on intended rankings, Vana and Lambrecht [10] show that the position in a list significantly influences user behavior even if the order is quasi-random and has not been sorted by the platform according to relevance. For data of an online retailer, they show that the product review at the top of the list has a significantly higher influence on the purchase decision than the product reviews listed below on the same page.

Jacobs and Hillert [11] show that the relevance of positions within a list also applies to security trading. Companies with a name of high alphabetical rank experience more trading volume and liquidity. The authors also find higher fund inflows for mutual funds

---

[2] https://www.zulutrade.com.

with names of higher alphabetical rank. This effect is especially strong for small mutual funds.

2.3 Rankings in the Context of Social Trading Platforms

Social trading platforms offer comprehensive screening instruments to increase transparency between traders and investors and therefore reduce information asymmetry. This could reduce the risk and adverse selection of investors. The ranking offered by the platforms is, besides others, one instrument to create transparency [5]. Röder and Walter [1] found that the 25 top-ranked portfolios on a social trading platform receive significantly higher fund inflows. They argue that social trading platforms can redirect inflows by adjusting the ranking's rules. This is in line with the findings of the literature on rankings in general. A similar relationship can be observed for investment funds. A good rating by Morningstar, one of the most popular agencies for ratings of mutual funds, leads to significantly higher capital inflows for these mutual funds [12]. Lee and Ma [13] criticize rankings on social trading platforms if they solely rely on past performance because it might result in risky behavior of traders and thus, ultimately increase investors' risk. To overcome this issue, they propose their own ranking system building on performance, risk, and consistency. Pan et al. [14] examined two measures (number of followers and past performance) that investors might use to rank the traders. A trading strategy, utilizing the number of followers to rank on, achieves significant positive returns if not more than the top 10 traders are followed. Utilizing the performance-based ranking leads to a positive return when investing in the top 50 traders or less. Strategies that invest in a broader base of top-ranked traders do not result in significant positive returns. Therefore, Pan et al. [14] conclude that there are only very few skilled traders on social trading platforms. Glaser and Risius [6] conclude that the choice of the measure for trader evaluation influences their trading behavior. The signal providers tend to adjust their behavior to optimize their own raking in correspondence with the present evaluation criterion.

In summary, it can be said that the rankings have a clear influence on investor behavior, but the economic benefits of these rankings for the investors have not been sufficiently analyzed. Even though Pan et al. [14] show for the platform eToro which benefits could arise from a ranking sorting on the number of followers or past performance, neither of these measures is the default option by which portfolios are presented to the user. In its current version, eToro recommends different traders to the user based on an internal recommendation system not transparent to the user. Only on the bottom of the page, traders are sorted by numbers of followers and past performance.

The platform ZuluTrade developed the ranking system called ZuluRank that ranks the traders on multiple parameters. Among these parameters are maturity, exposure and drawdown of the strategy implemented by the trader. However ZuluTrade does not reveal the complete set of parameters and the formula for the ranking system [15]. Furthermore, the default sorting option for traders in the current version of ZuluTrade is not ZuluRank but the profit made by investors copying the trader within the last month.

Against the theoretical background and to answer our two research questions, we formulate four different hypotheses, where the first two are related to **RQ1** and the second two belong to **RQ2**. To provide economic value to investors on social trading platforms, two conditions have to be fulfilled. First, investors on social trading platforms have to use the ranking at least to some extent when they make investment decisions. This condition leads to **H1a**. The extended literature on the effect of platform rankings already provides a strong indication for **H1a**. Furthermore, Röder and Walter [1] confirm this hypothesis. In order to validate our dataset and to deepen the insights of previous studies, we need to verify this hypothesis based on our data.

H1a: *A higher position within the portfolio ranking will result in higher future capital inflows for this portfolio.*

The second condition is met if the portfolio ranking is related to future portfolio returns. We derive **H1b** from Martens [7] who points out the advantages of platforms in terms of data collection and data analytics over the individual platform user. Therefore, we assume that the platform can identify portfolios of skilled traders that will outperform portfolios of non-skilled traders in the long run.

H1b: *Top-ranked portfolios outperform portfolios not listed among the top-ranked group.*

In order to answer **RQ2** and to investigate the underlying structure of the ranking, which may lead to an economic benefit of the ranking, we formulate the hypotheses **H2a** and **H2b**. We suspect that portfolios that have been in the top-ranked group prior to a crisis will follow a less risky approach during the crisis than portfolios not among the top-ranked. Kollock [16] argues that reputation systems on e-commerce platforms reduce the risk for platform users. Two reasons are given for this. On the one hand, those market participants with a very poor reputation are sorted out by the market mechanism or the platform provider. On the other hand, those with a good reputation have a high interest in maintaining it. As both mechanisms might also apply to the ranking of social trading platforms, we formulate **H2a** as follows:

H2a: *Top-ranked portfolios take a more defensive approach under high volatile market conditions compared to not top-ranked portfolios.*

The ranking of an online platform is not fixed but changes dynamically. To get a better understanding of the crash induced changes within the ranking, we formulate **H2b**. Since the rankings on social trading platforms typically include past performance as an evaluation criterion, a momentum strategy is implicitly pursued. If an investment strategy is dynamically-adapted to the current ranking, past winners are regularly bought and past losers are sold. This should lead to a strong procyclical strategy as stated in **H2b**, which should be particularly evident during the crash.

H2b: *Changes within the ranking during times of high volatility will result in procyclical investment behavior.*

3   Dataset and Descriptive Statistics

For this study, we utilized a hand-collected dataset from a major social trading platform focusing on equities. The platform's remuneration scheme is based on past profits and investment volume. We collected the daily top 100 ranked portfolios over almost five months, ranging from 12-10-2019 to 05-05-2020. This ranking is the default sort option in which the platform presents the portfolios to the user when searching for them. The ranking is based on different criteria, which are transparently revealed to the user along with the calculation formula used for the ranking. Among the criteria are portfolio age, past performance, trader activity, risk, and invested capital.

We further collected data of 1,000 randomly selected portfolios as a control group. To avoid a biased control group resulting from inactive or non-money carrying portfolios, we only sampled over active portfolios (traders have been active within the last two weeks before the observation start), which contain a minimum of 5,000 € invested capital. Portfolios that are constituents of the top 100 portfolios at the beginning of our observation period are not part of the control group. However, portfolios that became a part of the top 100 ranking later during the observation period are not excluded from the control group. This procedure also prevents a biased control group. Otherwise, portfolios that perform well in the future would be systematically excluded from the control group. In addition, we collected data on the portfolio composition as well as on the performance of each observed portfolio during the observation period.

4   Analysis

As already mentioned in Sect. 2, an analysis of the ranking's effectiveness is particularly important if investors rely on it when forming their investment decisions. The fact that a ranking influences the decisions of platform users is widely documented in the platform literature [8, 9] as well as in the literature on social trading [1].

To make sure that the ranking of the platform we are investigating also has this property, we test it on our data. To do so, we relate the platform provided ranking of each day with the inflows of the following day. Since the platform does not show inflows and outflows separately, only net inflows are considered. For each day, we rank the portfolios according to the net inflows they received on that day. This procedure eliminates the trend in the overall development of inflows, as the portfolios had suffered from a considerable overall withdrawal of capital during the beginning of the Covid-19 pandemic. The capital was later returned to the portfolios during the market recovery. We perform a ranking based on €-value net inflows as well as on relative net inflows. To calculate relative net inflows, the €-value of net inflows is related to the capital already invested in the portfolio. We calculate Pearson and Spearman correlation between the daily portfolio position within the platform ranking and the inflow-based ranking. We perform this calculation for the top 100 ranked portfolios, as well as for different subgroups of these 100 top-ranked portfolios. The results are presented in Table 1. When considering €-value inflows, the Pearson correlation amounts to 0.0776. A position at the top of the ranking is therefore accompanied by higher capital inflows during the next day. Using relative inflows leads to similar results (0.0752).

Table 1. Relationship between ranked inflows and platform provided portfolio rankings for different groups.

Ranking group	Correlation between platform ranking position and next day inflow ranking			
	€-value inflows		Relative inflows	
	Pearson	Spearman	Pearson	Spearman
Top 1–100	0.0776***	0.0754***	0.0752***	0.0702***
Top 1–20	0.1744***	0.1817***	0.1688***	0.1721***
Top 21–40	0.0005	0.0026	−0.0007	0.0007
Top 41–60	0.0175	0.0177	0.0152	0.0161
Top 61–80	0.0101	0.0058	0.0034	0.0018
Top 81–100	−0.0090	−0.0075	−0.0080	−0.0060

*p < 0.05/**p < 0.01/***p < 0.001

The top 100 ranked portfolios are further divided into blocks of 20 portfolios each. For each subgroup, the correlation between the position and the corresponding inflows is calculated. As Table 1 shows, only for the 20 highest ranked portfolios a significant correlation between platform ranking and inflows can be observed. For all other groups, the relative position within the ranking seems to be of no relevance. Applying Spearman as correlation measure leads to almost identical results. Our finding is in line with findings from other platform rankings that show a non-linear but convex relationship between ranking position and consumer behavior, for click-rates as well as purchase decisions [9]. It also provides evidence that tournaments that have been found in the mutual fund industry might also be an issue in social trading. The observed relationship between ranking and net inflows also provides evidence for certain trust the investors have in the platform's analysis capabilities. This is in line with the findings of Balyuk and Davydenko [2] on crowdlending platforms, according to which investors rely almost exclusively on the platform's analytical capabilities.

Our data support **H1a**. Nevertheless, the endogeneity of rankings is a fundamental problem when evaluating their causal effect on user behavior [8]. Our approach of using the net inflows from the day after the ranking at least excludes the possibility of inflows influencing the ranking itself. However, an omitted variable may drive both, the ranking as well as the inflows. Although, the available data do not allow for a more detailed examination of causality.

To evaluate the economic benefit an investor might gain from following the platform's investment recommendation expressed in the rating, we derive different trading strategies. The first strategy invests in an equally weighted portfolio containing the 100 highest-ranked portfolios according to the ranking on the day our observation starts (12-10-2019). We refer to this strategy as *top 100 [stable]*, which is represented in Fig. 1 by the dotted line. The second strategy also invests in the top 100 portfolios but adapts the portfolio according to the changes in the ranking over time. Changes in the ranking are checked daily during trading hours. If such a change is determined, the change is adjusted in the portfolio using the closing price. The performance of this strategy is shown as a dashed line in Fig. 1 and is called *top 100 [dynamically-adapted]* in the following.

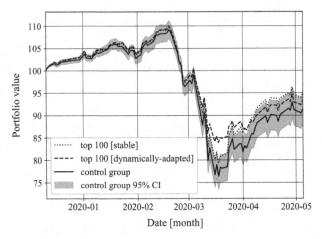

Fig. 1. Performance of equally-weighted top 100 strategies compared to the control group.

We consider a purely random portfolio selection to be the benchmark for these strategies. This is based on the control group of 1,000 portfolios and is shown as a solid line in Fig. 1. The two-sided 95% confidence interval of these strategies is based on a bootstrapping approach, where a total of 10,000 times portfolios consisting of 100 social trading portfolios each were drawn from the control group. All strategies are designed to remain equally-weighted over time. This requires daily rebalancing to offset weighting shifts resulting from differences in performance between the social trading portfolios. However, we do not account for any transaction cost that results especially from daily rebalancing.

Figure 1 shows that both ranking-based strategies generate a higher return than the control group during the observation period. However, at the observation period's end only the top 100 [stable] strategy is outside the control group's confidence interval. Before the market collapse started at the end of February 2020, the two top 100 strategies were very close and well within the benchmark's confidence interval. Only during the crash the ranking-based strategies performed better than the control group. It is particularly noticeable that a dynamic adaption according to the current ranking helps to significantly limit the downward potential. The low point of the strategy with continuous adjustments is a total of 2.91 (7.55) percentage points higher than the first strategy (benchmark). At the same time, the top 100 [dynamically-adapted] strategy benefits much less from the stock market recovery following the crash. Over the whole observation period of almost five months, the constituents from the top 100 [stable] strategy lost on average 4.81% of the capital invested. The loss generated by the control group was 8.57%. Utilizing the bootstrapping approach confirms that this difference between the two groups is significant ($p < 0.05$). This evidence supports H1b. Thus, the ranking on the social trading platform we analyzed can provide economic benefit to its platform users (investors). It can also be seen that a dynamic adaption according to the current ranking can reduce downward risk but also limits the upward potential following a market crash. As we cannot observe significant differences in the strategy performance prior to the market collapse, it seems more likely that the ranking can identify the portfolios with a successful market timing than portfolios with a successful stock selection.

The previous analysis showed that the strategies differed particularly in terms of their risk exposure. As our analysis is specifically aimed at analyzing the effect of the ranking in turbulent market phases, a more in-depth analysis of risk measures is required. So far, we have looked at portfolios, which themselves consisted of 100 social trading portfolios. This approach results in diversification of almost all idiosyncratic risk. However, it seems unlikely that a typical investor on a social trading platform would invest in 100 social trading portfolios simultaneously. Thus, in addition to the systemic risk, which an investor would bear anyway and which is of particular importance in the market situation considered here, an investor may also bear considerable idiosyncratic risks, if a trader's portfolio is not sufficiently diversified. To capture both, idiosyncratic and systematic risk, we calculate the risk measures at the individual portfolio level. Maximum relative drawdown (MRD) and volatility are utilized as risk measures. The MRD is an asymmetric risk measure, which describes the asset's maximum percentage drop during a given period and therefore captures the worst possible investment outcome during this period [17]. The distribution of MRD across portfolios is shown in the left histogram of Fig. 2. The annualized volatility is shown in the right histogram. Only the constituents of the top 100 [stable] strategy are included and are followed over the entire observation period. Adding the portfolios that later joined the top 100 would result in biased results, as the portfolios which later joined the top 100 ranking have logically performed well in the preliminary period before they joined.

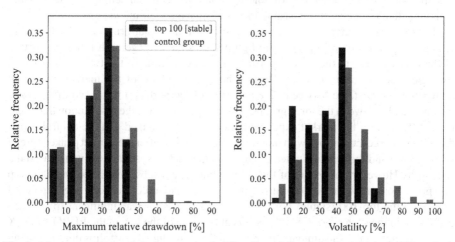

Fig. 2. Maximum relative drawdown and volatility of portfolios from the initial top 100 ranked portfolios and control group.

For both risk measures, a less risky approach can be observed for the top-ranked portfolios. The mean MRD of the top 100 ranked (control) portfolios amount to 27.40%

(30.44%). The difference between the two groups is significant ($p < 0.05$). While none of the ranked portfolios has an MRD greater than 50%, 7% of the portfolios from the control group lost more than 50% of the highest value previously achieved during the observation period. The results remain stable when utilizing the volatility as risk measure. The top 100 ranked portfolios have average volatility of 35.26%, which is significantly ($p < 0.05$) below the control group's volatility (41.85%). Among the top-ranked portfolios are also very few portfolios with extremely high volatility. Hence, a look at the risk measures shows that an investment in a top-ranked portfolio involves less risk than in a randomly selected portfolio. This provides the first evidence supporting **H2a**.

At this point, one could argue that the ranking, which also utilizes risk measures, favors generally less risky portfolios and those with lower market exposure and thus, naturally have a lower risk during a crash. However, this assumption is contradicted by the fact that during the bull market at the beginning of our observation phase, the top 100 strategies even performed slightly better than the control group. We calculate the portfolios' beta using the EURO STOXX 50 as market index during this bull market (observation start until EURO STOXX 50 peak) and could not find any differences between the beta of the top 100 [stable] portfolios and the beta of portfolios from the control group ($p = 0.9777$). However, for the following period (EURO STOXX 50 peak until observation end) we find significantly ($p < 0.05$) lower beta for the top 100 [stable] portfolios compared to the control group. This observation suggests that the groups follow different market timing approaches, but not that the top-ranked portfolios follow a less risky approach in general. There are different possibilities to conduct market timing and to reduce the portfolio's market exposure. This could be done for instance by using derivatives or by replacing high beta stocks with low beta stocks. However, a straightforward approach would be to sell the risky stocks and place the funds in the risk-free asset. As the possibility of holding cash is given in our setting, it is reasonable to assume that traders will use cash holdings to manage their market exposure.

Figure 3 shows that the traders on the social trading platform are indeed making strong use of this control option. The figure shows the average cash ratio (proportion of the cash position to the overall portfolio value) over time of the two ranking-based strategies and the control group. During the upward phase of the market, the mean cash ratio is very stable between 12% and 18% of the portfolio value and differs only insignificantly between the three groups. This further confirms the assumption that there is no difference in the market exposure between the top-ranked and the control group during this market conditions with moderate volatility. Just after the market peaked (green vertical line) and the market started to collapse, the traders systematically increased the cash ratio. However, the extent of the adjustment differs substantially between the top 100 portfolios and those of the control group. A comparison of the control group (solid line) with the top 100 [stable] portfolios (dotted line) shows a significant ($p < 0.05$) difference in the cash ratio of 8.00% points at the time of the low in the EURO STOXX 50 (red vertical line). In summary, investing in the top-ranked portfolios does not mean that one necessarily invests in a portfolio with lower market exposure, but that the top-ranked portfolios react more strongly to market changes and try to time the market as well as risk. Hereby **H2a** is confirmed.

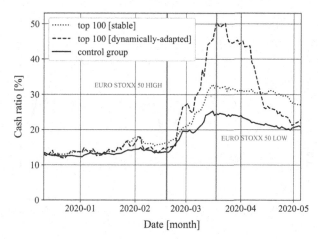

Fig. 3. Cash holdings of the top 100 strategies compared to the control group. (Color figure online)

It should be noted that this effect is primarily generated by an intended risk reduction of the traders and only results to a small extent from the crash induced value reduction of the risky portfolio stake. If we control for this value effect by considering the €-value cash positions instead of relative cash positions, the results' basic structure remains unchanged. Neither can the observed effect be explained by individual portfolios that have become inactive during the observation period.

The cash ratio of the strategy following the top 100 ranking dynamically (dashed line) shows that the ranking itself leveraged the change in cash ratio due to ranking adjustments. During the crash, portfolios with a low cash ratio are replaced by portfolios with high cash ratios. Shortly after the EURO STOXX 50 bottomed out, the ranking switches back to portfolios with higher market exposure. The observed changes in the ranking indicate a clear procyclical strategy and thus provide evidence for **H2b**. On the one hand, this procyclical adjustment limits the downward potential, on the other hand, this results in insufficient participation from the market's recovery phase. Both can be observed from Fig. 1.

So far, we have focused on the rankings capability to identify portfolios of skilled traders. However, the traders are also aware that they are ranked by the platform. We already showed that the ranking could incentivize tournament behavior. To evaluate this, it is necessary to control whether a trader's risk-taking depends on the position within the ranking. However, we cannot simply compare the risk of portfolios of different ranking levels, as the ranking is endogenous and depends also on the risk of the portfolios themselves. To avoid this problem, we observe the risk adjustment conducted by a trader after reaching (losing) a lucrative ranking position. Based on the convex incentive structure, we assume that a trader reduces (increases) her risk position if she reached (loses) a top position. We use changes in cash ratio as a measure for risk adjustments. Following an event study approach, we control for overall trends in the cash ratio by calculating the abnormal changes in cash ratio. The top group is defined as the top 10 ranked portfolios. In total, there are 100 changes between 36 different portfolios during the observation period.

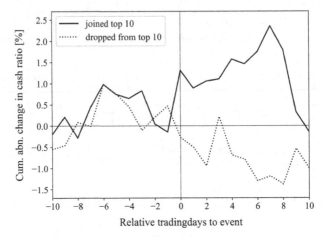

Fig. 4. Cash ratio event study on ranking changes.

Figure 4 suggests that the incentive structure induced by the ranking causes tournament behavior among traders. While the cumulative abnormal cash ratio before the ranking change remains relatively stable and close to zero, traders reduce the risk by increasing their cash ratio when they joined the top 10 (solid line). Traders that have dropped out of the Top 10 group show the reversed behavior. However, the observed difference between the two groups is insignificant. Therefore, the tournament cannot be confirmed based on the data at hand. Furthermore, these differences in the cumulative abnormal cash ratio do not persist over time but converge again after about 9 trading days. Even if this result cannot yet make a definitive statement, it does indicate an interesting direction for future research.

5 Discussion

Our study provides new insights into the ranking of social trading platforms. We take advantage of the first global stock market crash after the introduction of social trading platforms to test the use of rankings under market conditions where investors' capital is exposed to high risk. Our results contribute to a better understanding of the inner structures of social trading platforms as well as the role of rankings as a tool to reduce information asymmetries in general. It further relates to the literature that emphasizes the role of platform providers on FinTech platforms [2]. Our results show that rankings influence the investment behavior of investors (**H1a**). However, only for the top 20 portfolios within the ranking, the relative position within the ranking is important for future capital inflows. For the other portfolios, we do not observe any relevance of the relative ranking position. These results confirm the findings of previous research on platform rankings in general [8, 9]. Moreover, these results indicate that the mutual fund tournament also plays a role in social trading.

Based on trading strategies that utilize the platform provided rankings, we show that these rankings help investors to significantly increase performance (**H1b**) and reduce risk in times of crisis (**H2a**). These results indicate that the platform can identify portfolios of traders that are more successful in market timing and adjust their strategies more dynamically. However, there is no evidence that the platform can also identify those traders who can implement a more successful stock selection. Our results also show that traders actively switch between the cash position as a risk-free instrument and the risky investment in securities to manage their risk. The magnitude of the change in cash ratio we observed for social trading portfolios becomes clear when comparing it with the cash ratio change of mutual funds. The cash ratio of mutual funds increased from 4% in February 2020 to a mere 5.9% in April 2020 [18]. This shows how dynamic traders on social trading platforms can act. Investment funds are subject to much stricter regulations and would have to expect a considerable market impact if they were to sell their equity holdings in the way traders on social trading platforms did it. Market impact is still less of a problem for traders on social trading platforms, given the relatively low investment volume. The analysis of the time-varying composition of the top 100 ranked portfolios provides strong evidence for a procyclical behavior (**H2b**).

Finally, our first results of ranking-induced behavioral changes show a promising path for future research. A method based on the classical event study design is proposed to avoid the endogeneity problem of rankings. Future research could link the literature on tournaments with the social trading literature.

Our research has implications for various interest groups in the field of social trading and beyond. Active and potential investors on social trading platforms can benefit from the results to the extent that they can incorporate the ranking offered by the platform more consciously into their decision-making process. Platform operators can derive from our research that their ranking has a significant influence on the behavior of their investors and potentially also their traders. This also means that platform operators have a high level of responsibility when designing the ranking. Only the very few portfolios on top of the ranking are benefiting from it. This leads to high capital concentration among a few portfolios. Platform operators can use our results to evaluate how they can optimize the ranking to increase the added value for the investors and at the same time spread the capital over more portfolios to encourage more users to act as signal providers themselves. Furthermore, the work has implications for policymakers and regulators. Since the ranking is a tool to control the capital allocation on the platform, transparency towards the platform users is important. This transparency is very pronounced on the platform under consideration. However, this is not the case with all social trading platforms. Ultimately, our work provides the impetus for further research on rankings on social trading platforms and on their impact on changes in the behavior of traders.

The results of our investigation are subject to some limitations. The data utilized for this study are from a single social trading platform limiting the generalizability of the results. If other platforms do not offer a ranking or design the ranking according to different criteria, the results are only transferable to a limited extent. As we have shown in Sect. 2.3, the rankings of the different platforms cannot be compared directly, as not all platforms disclose details regarding their ranking computation. However, also other platforms like ZuluTrade include risk-based measures in their ranking systems [15]. Furthermore, we only consider a short observation period of almost five months. This allows us to capture very interesting properties of the ranking during a market phase of extreme volatility. At the same time, the results may not be applicable to other, particularly calmer, market phases. Besides, the properties of the ranking have to be evaluated on multiple market collapses to increase generalizability. However, as the market crash in connection with the Covid-19 pandemic was the first to this extent after social trading has been introduced, further opportunities to test this will only arise in the future. The benchmarks and trading strategies presented in this paper require daily adjustments. This would result in high transaction costs, especially if only small volumes are invested. However, we have not taken transaction costs into account in our modeling, as the strategies are not developed to trade on, but to implicitly show the ranking's benefit. As transaction costs would affect the ranking-based strategies and the benchmark simultaneously, the key result of our analysis would remain the same. Whereas, for the strategy that dynamically follows daily ranking changes, the transaction costs would be higher.

We have used the cash ratio to measure market exposure. As already mentioned, traders could also manage their market exposure without changing the cash ratio. Measuring market exposure via beta would cover all possible steering options. However, our approach by using the cash ratio allows us to measure market exposure on a daily basis, whereas beta can only be measured for a certain period of time. In addition, our data indicates that the cash ratio is the main steering instrument used by the traders. Thus, measuring market exposure via the cash ratio has a clear advantage for our application.

As was mentioned before, endogeneity is a fundamental problem when evaluating the effects of rankings on user behavior [8]. Since we cannot eliminate this problem based on the data available, it limits the results on **H1a** to a certain extend.

6   Conclusion

Our analysis shows that the ranking of portfolios published by social trading platforms actually provides valuable information for investors on these platforms. We look at the top 100 highest ranked portfolios of a social trading platform and compare them with randomly selected portfolios. The benefit is particularly present during turbulent market phases as we have experienced during the beginning of the Covid-19 pandemic. The ranking effectively identifies those portfolios with more successful management of market exposure and risk. Our data, however, do not show any indication that the ranking recognizes portfolios of traders who make a more successful stock selection. Thanks to successful market timing, the ranked portfolios were able to outperform their control group. Our results also show that investors can reduce their risk by investing in top-ranked portfolios. This protection mechanism is particularly strong when investors update their own portfolio according to the contemporary ranking. Such a dynamically-adapted strategy, however, carries the risk of procyclical behavior, which in turn can lead to reduced participation in market recoveries. Finally, our analysis shows that rankings also influence the behavior of platform users. Top-ranked portfolios gain significantly higher capital inflows than lower-ranked portfolios. Initial results also indicate that the traders evaluated by the ranking also react to it and adjust their behavior accordingly. This may result in tournament behavior, which has already been observed for mutual funds. However, to confirm or falsify this for social trading platforms as well, further analysis is required.

References

1. Röder, F., Walter, A.: What drives investment flows into social trading portfolios? J. Financ. Res. **42**(2), 383–411 (2019)
2. Balyuk, T., Davydenko, S.A.: Reintermediation in FinTech: evidence from online lending. In: Michael J. Brennan Irish Finance Working Paper Series Research Paper No. 18–17; 31st Australasian Finance and Banking Conference 2018. Available at SSRN 3189236 (2019)
3. eToro: Our Story. https://www.etoro.com/about/. Accessed 15 July 2020
4. Gomber, P., Koch, J.-A., Siering, M.: Digital Finance and FinTech: current research and future research directions. J. Bus. Econ. **87**(5), 537–580 (2017). https://doi.org/10.1007/s11 573-017-0852-x
5. Doering, P., Neumann, S., Paul, S.: A primer on social trading networks – institutional aspects and empirical evidence. Presented at EFMA Annual Meetings 2015, Breukelen/Amsterdam. Available at SSRN: 2291421 (2015)
6. Glaser, F., Risius, M.: Effects of transparency: analyzing social biases on trader performance in social trading. J. Inf. Technol. **33**(1), 19–30 (2018)
7. Martens, B.: An Economic Policy Perspective on Online Platforms. Institute for Prospective Technological Studies Digital Economy Working Paper 2016/05 (2016)
8. Ursu, R.M.: The power of rankings: quantifying the effect of rankings on online consumer search and purchase decisions. Market. Sci. **37**(4), 530–552 (2018)
9. Ghose, A., Ipeirotis, P.G., Li, B.: Examining the impact of ranking on consumer behavior and search engine revenue. Manage. Sci. **60**(7), 1632–1654 (2014)
10. Vana, P., Lambrecht, A.: The Effect of Individual Online Reviews on Purchase Likelihood. Tuck School of Business Working Paper No. 3108086. Available at SSRN 3108086 (2018)
11. Jacobs, H., Hillert, A.: Alphabetic bias, investor recognition, and trading behavior. Rev. Finan. **20**(2), 693–723 (2015)
12. Guercio, D.D., Tkac, P.A.: Star power: the effect of morningstar ratings on mutual fund flow. J. Finan. Quant. Anal. **43**(4), 907–936 (2008)
13. Lee, W., Ma, Q.: Whom to follow on social trading services? A system to support discovering expert traders. In: Tenth International Conference on Digital Information Management (ICDIM), Jeju Island, South Korea, pp. 188–193 (2015)
14. Pan, W., Altshuler, Y., Pentland, A.: Decoding social influence and the wisdom of the crowd in financial trading network. In: Proceedings of the International Conference on Privacy, Security, Risk and Trust (PASSAT), Amsterdam, Netherlands, pp. 203–209 (2012)
15. ZuluTrade: ZuluRanking system. https://www.zulutrade.com/zuluranking. Accessed 15 July 2020
16. Kollock, P.: The production of trust in online markets. In: Lawler, E.J., Macy, M., Thyne, S., Walker, H.A. (eds.) Advances in Group Processes, vol. 16, pp. 99–123. JAI Press, Greenwich, CT (1999)
17. Vecer, J.: Preventing portfolio losses by hedging maximum drawdown. Wilmott **5**(4), 1–8 (2007)
18. Ksenia, G.: BofA Poll Shows 'Extreme' Investor Pessimism With Cash at 9/11 High. https://www.bloomberg.com/news/articles/2020-04-14/bofa-poll-shows-extreme-inv estor-gloom-with-cash-at-9-11-high. Accessed 15 July 2020

What Do Robo-Advisors Recommend? - An Analysis of Portfolio Structure, Performance and Risk

Albert Torno[✉] and Sören Schildmann

Chair of Electronic Finance and Digital Markets, University of Goettingen, Goettingen, Germany
albert.torno@uni-goettingen.de,
soeren.schildmann@stud.uni-goettingen.de

Abstract. Robo-Advisors guide investors through an automated investment advisory process, recommend personalized portfolio assignments based on their individual risk-affinity as well as investment goals and rebalance the portfolio automatically over time. Giving basic investment advice to customers, it can provide a useful way to reduce risk by diversifying and mitigating biases, while keeping a certain degree of performance at low costs. To verify these claims we conduct a sophisticated analysis of recommended portfolios of 36 Robo-Advisors, based on six distinct model customers with different risk-affinities and investment horizons, resulting in 216 recommended portfolios. We find that the analyzed Robo-Advisors provide distinct recommended portfolios for the different risk/investment horizon combinations, while sharing similarities in used products for portfolio allocation. We also find issues within the recommended portfolios, e.g. a low degree of distinctiveness between different investment horizons and a high amount of equities even in the short-term investment horizon.

Keywords: Robo-Advisor · FinTech · Portfolio analysis · Performance and risk analysis

1 Introduction

The Digital Transformation of industries lead to enhancements of operational process as well as changes in customer experiences and business models. Within the financial industry incumbent financial firms, as well as new start-ups, so called FinTechs, are competing in digitalization and automatization efforts to provide new ways of customer service, for example in wealth and asset management [1]. Traditionally, wealth and asset management are based on human interactions and trust between a financial advisor and the customer, which is a time consuming and costly process. This business model requires high capital investments to be profitable for financial advisors. Therefore, financial advisors mostly offer these services to high-net-worth individuals [2]. To provide financial advisory services to a wider range of customers and reduce costs, the Digital Transformation lead to the development of so called Robo-Advisors (RA) [3, 4]. A RA is an information system (IS), which guides investors through an automated investment

© Springer Nature Switzerland AG 2020
B. Clapham and J.-A. Koch (Eds.): FinanceCom 2020, LNBIP 401, pp. 92–108, 2020.
https://doi.org/10.1007/978-3-030-64466-6_6

advisory process, recommend personalized investment portfolio assignments, based on their individual risk-affinity as well as investment goals and rebalances the portfolio automatically over time [2, 4]. Low interest rates, as well as new und cost effective financial products like Exchange Traded Funds (ETFs) are the basis of an upcoming customer interest in the financial markets, even with low amounts of invested capital [4]. RA providers claim to simplify investment decisions and financial planning, but the advice provided is considerably less comprehensive than that of a human financial advisor [5]. Nevertheless, especially first-time investors can benefit from basic investment advice to mitigate basic portfolio mistakes, like low diversification, home bias or other human tendencies researched within the domain behavioral finance [6].

While research on RA portfolio structures is still rare, lacking the analysis of a sufficient sample of portfolio structures as well as risks and performances, RA comparison websites emerged, e.g. [7, 8], that try to provide recommendation for the best RA to potential customers. However, these websites only consider and track one portfolio per RA and are just focusing on performance, without the relation or analysis of the associated risks. Also, the websites do not disclose which underlying methods or customer characteristics are used within the RA recommendation procedures. This paper contributes to a better understanding of RAs by systematically capturing and analyzing these recommended portfolios and thus answers the following research question:

RQ: Which similarities and differences do the portfolios recommended by RA have, regarding portfolio structure and selected products, performance and risk?

We examine 36 RAs and analyze a sample of 216 distinct portfolio recommendations, based on six distinct model customers with low, medium and high risk-affinities, as well as different investment horizons of 3 and 15 years. Based on our analysis using various performance and risk indicators, we provide novel insights on recommended portfolio structures of RA for research and practice.

To answer the research question the remainder of this paper is structured as follows: Starting with the theoretical foundations we describe the key characteristics and process phases of RA, as well as present existing research on RA portfolio structures. Secondly, we describe the methodology by explaining the analysis approach as well as analysis measures and statistical test procedures. Thirdly, we present the findings of our analysis, followed by a discussion including implications for practice and research as well as providing limitations and future research directions. Finally, the conclusion summarizes the most important findings and implications.

2 Foundations and Related Research on Robo-Advisors

2.1 Robo-Advisor Definition and Process

The term *Robo-Advisor* is comprised of two components: "*Robo*" as an abbreviation of *robot* meaning "a machine controlled by a computer that is used to perform jobs automatically" and "*Advisor*" meaning "someone whose job is to give advice about a subject" [9]. A RA in the field of asset management can therefore be defined as an automated system, which undertakes the role of a financial advisor. Therefore, the RA providers' aim is to digitize and automate the entire traditional financial advisory and asset management process [2]. Instead of a human financial advisor analyzing the financial situation of a customer, a typical RA uses algorithms to combine customer information with a suitable portfolio recommendation [10]. As input, the RA uses the characteristics entered by the customer regarding her person, goals and risk-affinity [11]. Based on this information RAs recommend a personalized investment portfolio to the customer and rebalance it over time. The RA process can be divided into four phases: *Initiation, RAP, Matching and Maintenance* [2].

Initiation. In the first phase of the process, information asymmetries between providers and customers are dismantled [2]. Due to the absence of human interaction and the importance of trust in financial advisory, RA providers aim at enhancing transparency by giving information about the whole advice process, the products used, and the costs associated with the services. This transparency not only has a positive effect on the attitude of the customers to the RA provider, but also increases the customers willingness to accept costs [12]. In this phase, providers also make a pre-selection of products (mostly ETFs) from which their recommended portfolios are compiled. Besides the preselection of funds by the RA provider, the customer often is asked for her investment objective, which may have an influence on the following phases.

Risk Assessment Process (RAP). The formulation of an investment goal leads into the RAP. In this phase, the risk profile of the customer is created, which is the basis of the portfolio recommendation. In an online questionnaire the customer goes through an self-assessment of her risk-affinity and investment characteristics [13]. The questions asked can be divided into three types: (I) General information, (II) Risk capacity and (III) Risk tolerance [13]. It could be shown that RAs ask an average of ten questions, six of which have an impact on the risk profile [13]. Creating a risk profile using a mostly static online questionnaire is a rather simple procedure, in comparison to traditional financial advice. It implies that clients have only one single and static risk preference. The literature in this area questions, whether the actual risk affinity of the client can be derived with this method alone [6, 10]. RAs increasingly use more sophisticated methods to create a risk assessment e.g. by utilizing metaphors and scenario-based questions. RA thereby aiming at a balance between simplicity and sophistication of the risk profile [5].

Matching. Within the matching phase an algorithm transfers the answers provided into a risk profile of the customer and recommends a corresponding portfolio from the investment product space of the RA provider. Since matching is usually based on an unpublished algorithm, the exact way of how a risk profile is matched to a portfolio is considered a black box [2, 5]. However, an investigation of 219 RAs showed that that over 80% of RA providers base their recommendation according to their own statements on three methods: (*I*) Modern portfolio theory according to Markowitz, (*II*) Model portfolios and (*III*) Portfolios with constant weightings [10]. Thereby, it was found that more sound methods of portfolio creation also lead to a higher total Assets under Management (AuM) sum of the RA providers [10]. When presenting the recommended portfolio structure to the customer, some RA provide the option to modify her risk class and/or modify the portfolio structure in various ways. Logical risk verifications are run to ensure, that the customer does not change too much from the associated risk profile. [2]. Typically, the end of the matching phase is marked by an offer. If not already done to this point, the personal data of the customer is now required, and an account opening is initiated.

Maintenance. In the last phase of the process, rebalancing and reporting takes place. By means of automated rebalancing, the RA maintains the portfolio weights of the individual asset classes and thus ensures that the risk of the entire portfolio remains stable. By maintaining a desired portfolio risk, both over- or underperforming individual products, as well as reactions to external effects can lead to changes in the portfolio structure [14]. Each RA provider pursues an individual strategy, thereby using fixed time intervals (e.g. quarterly, yearly) and/or trigger events (market changes, customer changes) for starting a rebalancing action. Because the maintenance is delegated from a human financial advisor to an algorithm, consequences of behavioral finance e.g. irrational human behavior in financial markets and biases should be reduced [6]. Lastly, within the maintenance phase RA do reporting. Besides the permanent available online access, the RA providers offer their customers in differing intervals reporting functionality as well as financial news and occasionally educational content.

2.2 Research on Robo-Advisor Portfolio Structures

Research on RA portfolio structures is still in its infancy and often based on small samples of internal data provided by one RA provider or uses external data, with often non-transparent data capturing procedures and analysis methods.

Using internal data provided by one RA, [5] analyze the aspects of behavioral biases like return chasing and the disposition effect and find a significant increase of diversification in RA portfolios compared to portfolios managed by the investor themself. Also based on internal data of one RA, [15] concludes that a reduction of the minimum account size led to a net increase in total deposit inflows and an increase of new accounts by less-wealthy investors.

Using external accumulated data, [10] investigate portfolio recommendations in form of quantitative methods used inside the RA, by studying RA websites and white papers. This qualitative research was based on 219 existing RAs (reduced to 28 for the actual analysis), and shows a high usage of classic Modern Portfolio Theory within RA algorithms. [16] investigates which data about an investor should have a high influence on recommended portfolios and conclude that information about the ability to take risks (e.g., net wealth) and the willingness to take risks (e.g., risk aversion) should have the biggest impact. The panel study, while helpful for RA providers, didn't analyze how or what RA actually recommend within their portfolios. [11] conclude, that RA while asking many questions, do not use all information adequately to model recommended portfolios, thus concluding that "most RA provide generic and poorly individualized advice". [17] evaluate the investment performance of four RA for four investment horizons (1–3, 4–6, 7–15 and 16–40 years), but without capturing distinct portfolios with varying risks per RA.

3 Methodology

3.1 Analysis Approach

To answer the research questions, we started by developing distinct model customers, to ensure a neutral, transparent, and replicable analysis procedure. Firstly, we considered, which characteristics of a customer are highly impactful for the RA portfolio recommendation. Based on a first sample of five RAs and various combinations of differing response patterns and the work of [16], we choose three distinct risk and two distinct investment horizons for differentiation between recommended portfolios, thus representing the six distinct model customers shown in Table 1. These dynamic characteristics had the biggest impact on the differing recommendations of RA portfolios. While we already described the risk associated to a customer in our RA foundations chapter, we want to briefly state our rationale for the chosen investment horizon timeframes. In research and practice, short- and long-term investment horizons stand for differing investment behaviors and asset holding patterns [18, 19]. How many years of an investment horizon classifies as short-term or long-term is, to our knowledge, not defined in literature. We chose the short-term investment horizon to be three years, since e.g. a two-year investment horizon is to short for some RA to recommend a portfolio at all. We choose 15-years as the long-term investment horizon, because it is well above the three-year timeframe and therefore distinctly different to the short-term investment horizon.

Besides the dynamic characteristics, we choose static characteristics, that have not changed during the various RAP runs (e.g. age, sex, investable capital, savings ratio), to reduce complexity and ensure a manageable execution of RAP and subsequent analysis of the portfolios. All static characteristics had a low or none impact on the portfolio recommendation. The rationale behind each value of the static characteristic of the model customer, as well as exemplary questions and answers to model the static and dynamic characteristics are presented in Table 4 and Table 5 in the appendix.

Table 1. Risk/investment horizon combinations of model customers

(Lo3): Low risk / investment horizon 3 years	(Lo15): Low risk / investment horizon 15 years
(Me3): Medium risk / investment horizon 3 years	(Me15): Medium risk / investment horizon 15 years
(Hi3): High risk / investment horizon 3 years	(Hi15): High risk / investment horizon 15 years

After defining the model customers, we needed to find suitable RA to derive portfolio recommendations. The selection of RAs was based on literature references and an explorative internet research [8, 10]. Only if the RAP phase led to a portfolio recommendation without registering with personal data (e.g. social security number), the RA could be considered for our analysis. We thus obtained the following three categories of data availability of the RAs portfolio recommendations:

- *(Category A):* Fully transparent portfolio structure, including weightings per asset class as well as associated products.
- *(Category B)*: Semi-transparent portfolio structure with weightings per asset class, but no associated products.
- *(Category C)*: No transparency concerning portfolio structure.

RAs of category A provide the best data basis and allow a comprehensive analysis not only of the structure but also for risk and performance measures of the recommended portfolios, since our data analysis is based on historical prices of the products in the portfolios. These could only be gathered, if the names of the portfolio products could be obtained and matched with its International Securities Identification Number (ISIN). For these products historical daily closing prices in U.S. dollars between October 23rd, 2009 and October 25th, 2019 were retrieved from Thomson Reuters via DataStream. Based on that data, we could calculate annual averages of various performance and risk measures to simulate how the recommended portfolios would have performed in the given timeframe backwards from 2019 to 2009. RAs of category B allow at least the analysis of the portfolio structure, since they present a portfolio allocation at the end of the matching phase which can be used as a risk indicator [13].

A first descriptive analysis shows that 40% of the RAs present a transparent portfolio structure, including the used products. While 26% show at least the portfolio structure with asset classes but without products, 34% do not provide any information on where the capital will be invested before registration. As a result, we collected data from 20 RA providers of category A (ten from Germany, seven from the USA, two from the UK, one from Singapore). Two of these RA providers offer additionally topic-specific portfolio recommendations on sustainability (climate neutral investments) and gender diversity. These topic-specific portfolio recommendations were handled as separate RA in our analysis, as they were clearly distinguishable from the main RA recommended portfolios. In addition, we found and analyzed 13 RAs of category B (seven from Germany, four from the USA, one from Switzerland and the UK). Lastly, we found 16 RAs which fit to category C and are therefore not considered within our analysis. Table 6 provides an overview of the analyzed RA per category and country of residence.

For each of the considered 36 RAs in categories A and B, the RAP was passed through six times, matching the distinct risk/investment horizon combinations of our defined model customers. Each corresponding portfolio recommendations was captured in our database. Since the data collection took place at a certain time period (September 01st, 2019 until October 10th, 2019), the RA portfolio recommendations are a snapshot and may not be identical in other periods. Figure 1 summarizes our described analysis approach.

3.2 Analysis Measures and Statistical Test Procedures

Performance Measures. The historical returns of the RA recommended portfolios are serving as our main performance measure for the analysis. Therefore, based on the daily closing prices, steady daily returns of the products were calculated for available data points [20]. Afterwards, the simple return rate of the total period could be calculated per product. The average annual return on a product was calculated using the geometric

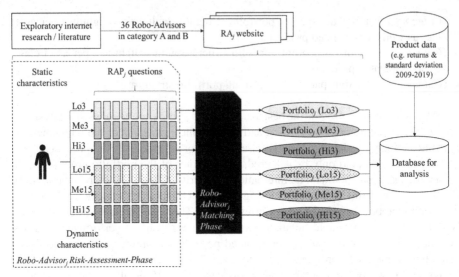

Fig. 1. Analysis approach with risk-assessment and RA recommended portfolio capture

return for the available period. This ensures, that compared to the arithmetic mean, the intra-year interest on simple returns is considered [21]. By adding up the N products within the RA recommended portfolios, weighted according to their portfolio share, the average historical portfolio return per year could be obtained [22].

Risk Measures. Within our analysis we used two indicators as a measure of risk: The standard deviation per year (volatility) and the Value at Risk (VaR). Firstly, to calculate the average annual standard deviation of a product, the daily standard deviation of the continuous daily returns was calculated for each product and afterwards annualized [21]. To calculate the standard deviation of an entire portfolio, a correlation matrix of daily returns of all considered products was created. Based on that matrix, we calculated the correlation coefficients of the individual assets contained in a portfolio [22]. These correlation coefficients are used to calculate the mean portfolio variance, as well as the mean portfolio standard deviation p.a. In addition to volatility, the annual VaR serves as another highly established risk indicator for our analysis. The VaR can be understood as the total loss of an investment position, which with a certain probability, will not be exceeded in a certain period of time [21]. The VaR p.a. was calculated for confidence levels of 95% and 99% using the variance-covariance method. Thereby, the normally distributed annual standard deviations of the portfolios serve as the basis for calculation.

Sharpe Ratio. The Sharpe Ratio (SR) is a key figure that relates performance to risk. Based on Markowitz [23] and Tobin's findings on efficient portfolios, the SR is suitable to describe the best portfolio on an "Efficiency Line" using only one key figure [24]. In our analysis, the SR was calculated for both products and portfolios and shows the excess return over the risk-free interest rate per unit of standard deviation [25]. The excess return can be calculated by subtracting the risk-free interest rate from the mean annual portfolio return and dividing by the mean standard deviation p.a.

Statistical Test Procedures. Because most of our variables do not meet the normal distribution requirement for parametric tests, two nonparametric Signed-Rank tests were used for our analysis. The *Mann-Whitney-U-Test* was used to investigate the relationship between two unrelated samples. It is the non-parametric alternative to the t-test and checks the distinction between the mean values of the distributions of groups [26]. The null hypothesis supports the equal distribution of the variable in the groups. Therefore, rejecting the null hypothesis significantly describes a non-random relation of the groups. The direction can be determined by the middle rank [26].

For k independent samples (groups) the *Kruskal-Wallis-H-Test* was used in conjunction with the *Dunn-Bonferroni-Test*. This corresponds to the single factor variance analysis for parametric data. Here, too, the correlations were tested for a statistically significant difference in the ranks of the groups formed in the 95% confidence interval. The first test shows whether the groups differ significantly from each other. If they differ, those groups between which the difference exists are determined by pairwise comparisons. The preservation of the significance level α in repeated pairwise comparisons is ensured by the Bonferroni method [26]. The effect strength r of the tests used can be approximated from the z value in SPSS, where $0 < r \leq 0.3$ describes a small, $0.3 < r \leq 0.5$ a medium and $r > 0.5$ a strong effect [27].

4 Findings

4.1 Portfolio Products and Allocation

A total of $N = 214$ different products are utilized by the analyzed RA in category A for their portfolio allocations, consisting of 143 ETFs (~2/3) and 71 (~1/3) mutual funds. The largest issuers of products are iShares by Blackrock with 25%, Vanguard Group with 11% and Xtrackers with 9%. The most frequently recommended products in the portfolios, measured by AuM, are Vanguard emerging markets ETF (VWO), the Vanguard FTSE developed markets ETF (VEA) and the iShares Core S&P 500 ETF (IVV). Figure 2 shows the return and risk (in form of standard deviation) box plots of all products, that are used within the RA recommended portfolios.

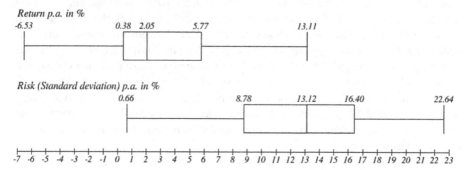

Fig. 2. Return and risk (standard deviation) box plots of all RA used products

After analyzing the products used by the RA in isolation, we investigated the portfolio allocations of all 216 RA recommended portfolios (RAs of category A and B). Table 2 shows the average, minimum and maximum weightings of the asset classes within the RA recommended portfolios in percent. The asset weights do not sum up to 100%, since some asset classes with low weights are left out.

Table 2. Portfolio allocation in % for each model customer

Asset class	Lo3			Me3			Hi3		
	Avg.	Min.	Max.	Avg.	Min.	Max.	Avg.	Min.	Max.
Equities	21.27	0	51	46.40	11	76	69.89	13	100
Cash	5.6	0	48	2.83	0	42	1.84	0	35
Gold	0.76	0	5	0.39	0	5	0.33	0	5
Commodities	0.65	0	7.5	1.3	0	6.1	1.45	0	12.5
Government Bonds	45.27	0	89	30.06	0	60	15.93	0	55
Corporate Bonds	25.38	0	80	17.36	0	36	8.94	0	30
Asset class	Lo15			Me15			Hi15		
	Avg.	Min.	Max.	Avg.	Min.	Max.	Avg.	Min.	Max.
Equities	25.01	0	79	53.14	30	84	80.84	39.59	100
Cash	4.33	0	48	1.55	0	10	0.82	0	6
Gold	0.76	0	5	0.39	0	6	0.33	0	5
Commodities	1.38	0	7.5	1.5	0	11	1.31	0	10
Government Bonds	42.79	0	89	25.33	0	60	7.68	0	33
Corporate Bonds	24.91	0	80	16.33	0	35	7.24	0	32

Since equity funds have a higher risk/return profile compared to the other asset classes, their portfolio weighting is often used as a risk-return indicator. We find that across all RA providers, the risk class determined in the RAP has a high impact on the recommendation of the RA. As the customer's risk affinity increases, the equity quota in the portfolio rises and the quotas of less volatile products, like government and corporate bonds, fall. With the change from the risk class low to medium, the equity ratio in the average portfolio increases by 100% for both investment horizons. When choosing a medium-risk instead of a high-risk profile, the result is an increase in the equity quota by 50%. The proportion of bonds decreases with the change from low-risk to medium-risk by about 30–40% and from medium-risk to high-risk by 50–70%. On average, the pairing of risk-affinity and portfolio recommendation regarding the asset class can be considered as working.

However, in terms of investment horizons, the three risk classes and their asset class ratios barely differ from each other: Major differences in the group averages are only noticeable between Hi3 and Hi15. The equity quota is 15% higher for the long-term portfolios, while the government bond quota is twice as high on average in the short-term portfolios at 15.90% vs. 7.68%.

The minima and maxima displayed in Table 3 show how different the RA recommended portfolios are. Especially, the Hi3 portfolios seem to vary highly between different RAs: For example, the equity ratios range between 13% and 100%. With an average annual VaR (95%) of 26%, equities have the highest volatility of the asset classes and therefore risk associated to them. In this context, e.g. UBS recommends an investment horizon of at least five years for its equity funds. Nevertheless, for a high-risk/performance portfolio, the recommendation of a high ratio of equities can probably be justified for a short-term investment horizon. However, recommending a maximum equity allocation of 51% (79% for 15 years) for a low-risk/performance class and an investment horizon of three years seems questionable. The big differences between RA recommended portfolios show that there is little consensus between the RA on what the best investment strategies and therefore portfolio allocations are.

4.2 Portfolio Performance and Risk

In the following we present our findings regarding portfolio performance and risk. As stated in the methodology, this in-depth portfolio analysis could only be done with the recommended portfolios the 23 category A RAs. Figure 3 presents the μ-σ-chart for the N = 138 portfolios of these RAs, with averages given for the six distinct model customers. To put the RA portfolios into relation, we also present three benchmark indices, that represent the asset classes equities (MSCI World), corporate bonds (Bloomberg Barclays Global Aggregate Corporate Bond) and government bonds (Bloomberg Barclays Global Treasury).

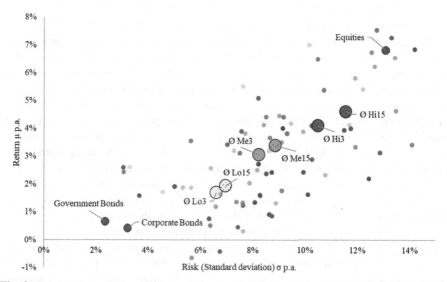

Fig. 3. RA recommended portfolios of category A in the μ-σ-chart with benchmark indices (October 2009 to October 2019)

Table 3. Portfolio analysis concerning risk classes

Kruskal-Wallis-H-Test	Risk class/Product	N	Mean values (left) and middle rank (right) for each variable					
Variable			Return p.a.		Standard deviation p.a.		Sharpe-Ratio p.a.	
RA Portfolios	Low	46	1.70%	38.52	6.35%	37.52	0.22	46.93
	Medium	46	3.22%	73.11	8.48%	65.63	0.35	77.63
	High	46	4.38%	96.87	11.08%	105.35	0.37	83.93
	Total	138						
Pairwise comparisons			Corr. Sig. by Bonferroni (2-sided) (95%)			Effect strength r		
Return p.a.	Lo-Me		$\rho < 0.001$*			0.433		
	Lo-Hi		$\rho < 0.001$*			0.730		
	Me-Hi		0.013*			0.297		
Standard deviation p.a.	Lo-Me		0.002*			0.352		
	Lo-Hi		$\rho < 0.001$*			0.848		
	Me-Hi		$\rho < 0.001$*			0.497		
Sharpe-Ratio p.a.	Lo-Me		$\rho < 0.001$*			0.384		
	Lo-Hi		$\rho < 0.001$*			0.463		
	Me-Hi		1.000			–		

It is noticeable, that there are major differences between the RA portfolio recommendations. Many portfolios exist that dominate others with the same risk and higher returns p.a. or vice versa. However, these differences can't be significantly grouped by the investment horizon of portfolios. On average, portfolios recommended for a term of three years have an annual return of 2.95% and a standard deviation p.a. of 8.43%. Portfolios with an investment horizon of 15 years have an annual return of 3.3% and a standard deviation p.a. of 9.1%. The statistical analysis of these two groups using the Mann-Whitney-U-test does not show significant differences for the variables return p.a., standard deviation p.a., VaR p.a. (95%) and SR p.a. Therefore, it can be stated, that investment horizons play a minor role in the recommendation of RA portfolios.

In contrast to the investment horizon groups Table 3 shows by utilizing the Kruskal-Wallis-H-Test, that the portfolio recommendations differ significantly for each risk class regarding their return p.a., standard deviation p.a. and partly SR p.a. As expected, the greatest difference can be observed in the risk/return combination from the lowest to the highest risk class. We also found a positive correlation between the mean values of the risk classes, which implies that a higher return is associated with a higher risk within the portfolio. Lastly, the group of the risk class "low" has significantly lower returns per unit of risk (measured by the SR p.a.), than the other two risk classes.

Summarizing our findings, it can be stated that a higher risk-affinity entered in the RAP does lead to the recommendation of a higher risk/performance portfolio. In contrast, a lower risk-affinity leads to significantly lower risk/performance portfolio recommended by the RAs. The RAP and matching phase of the analyzed RA therefore work as intended.

5 Discussion

5.1 Implications

The results of our study provide several important implications for practice and research. Our analysis shows that interested investors should not trust their preferred RA blindly. Especially when considering shorter investment horizons, investors should be aware that RA, despite of their high volatility, often use large shares of equities in their recommended portfolios. This may be due to the need of RA to offer clients a high return, but it also has the disadvantage, that if a crisis occurs and the investment horizon is exceeding, a loss must be realized.

RAs intend to offer solutions for their risk-averse investors with short-term investment horizons but providing such solutions within the current low-interest market can be complicated, and therefore too complex for RAs. The volatility of equities and bonds makes it at least questionable to invest in short-term time horizons using trading algorithms based on passive investments and diversification strategies, that are often used within RAs [2, 4]. To provide a more sophisticated financial advisory service, RAs need to consider more financial products with lower volatility for short-term investments like savings accounts. An alternative approach is to neglect short-term investment horizons entirely like Scalable Capital does, thus refusing to provide a portfolio recommendation for short-term investment horizons. This seems to be, when no other options provided, the best advice a RA could give to a short-term investor.

Finally, our analysis indicates, that the analyzed RA do not include the captured data sufficiently into their recommendation process, or a high amount of information does not change the recommended portfolio structure. Thus, we agree with [11] and state, that RA provider should use the data captured within the RAP for more individualized portfolio recommendations or improve the RAP in a way, that RAs just ask for information, that is actually used within their recommendations.

5.2 Limitations and Future Research

Despite the careful design of our study, this paper is subject to several limitations. Firstly, since our data collection took place at a certain time period between September and October 2019, the recommendations of the RAs are only a snapshot and may not be identical with recommendations from other periods. Within our analysis we did not capture the maintenance phase of the RA. Therefore, we do cannot provide insights on how often and to which degree the different RA are rebalancing their recommended portfolios. By analyzing the portfolios of the RA at various times, the changes within the portfolio recommendations could be observed and analyzed. Also, it can be of interest, how the RA approach the more volatile financial markets after the financial crash due to the Coronavirus-pandemic.

Another limitation regarding our analysis can be seen in the data collection from the perspective of six distinct customer models, having different risk-affinities and investment horizons. While carefully modelling the dynamic characteristics of our model customers and rigidly going through the RAPs, it is possible, that occasionally questions couldn't be answered with confidence according to the risk-affinity of the model customer. Also, more granular portfolio changes through more differentiating individual customer profiles, e.g. by varying our static characteristics, could be helpful to collect more varying RA recommended portfolios. This can lead to a broader understanding of the relationship between the RAP and RA recommended portfolios.

Lastly, the different monetary policies of the US Federal Reserve System (FED) and the European Central Bank (ECB) lead to separate market conditions in the country groups, which made comparisons difficult. This problem could be solved by considering different interest rates at least for the calculation of the SR p.a. Due to the lack of appropriate conditions for parametric tests, the non-parametric alternatives were used. These characterize in a lower test strength but are considered more robust. The results of the tests were clear and plausible, so there is no reason to assume, that the type of statistical test mislead our results.

6 Conclusion

Previous research on Robo-Advisors imply, that they provide basic investment advice and management. This paper aimed at providing new insights on the question of how RA recommended portfolios are structured, especially concerning performance and risk. Therefore, we analyzed a sample of 36 RA and 216 distinct recommended portfolios for six defined model customers, between September and October 2019.

The results of our study state several important implications for practice as well as for research. Firstly, we show that the basic investment advice of RA functions sufficiently, providing appropriate investment advice, at least for higher risk-affinities and long-term investment horizons. The recommendations of the various providers however vary greatly, especially for customers with a short-term investment horizon and high risk-affinity. When the investment horizon is getting into mid- or short-term territory, RA recommend on average very risky portfolio allocations for passively managed investment portfolios. Furthermore, we show that investment horizons play a subordinate role in the recommendation of a RA portfolio, despite its impact on the long-term return of especially volatile asset classes like equities.

Appendix

Table 4. Exemplary static characteristics of model customer

Exemplary questions	Static characteristics	Rationale
Gender and age	Male, 30	Target group of RA: Young, digital native and professionally successful individuals, without financial affinity and time to take care of finances themselves [4, 14]
Occupation	Employee	
Annual income	50,000–100,000 €	
Savings rate	7%	Corresponds to the average savings rate of households in 18 European countries, as well as the USA and Canada between 2016 and 2018 [28]
What amount would you like to invest?	42,500 €	Corresponds to the average AuM per user for RAs in the USA (approx. 81,000 €) [29] and Germany (approx. 4,000 €) [30] in 2019
What is your total wealth?	50,000–180,000 €	Corresponds to the net private financial assets per capita in Germany and the USA in 2018 [31]

Table 5. Exemplary questions concerning dynamic characteristics of model customer

Exemplary questions	Dynamic characteristics		
	Short-term investment		Long-term investment
Investment horizon?	3 years		15 years
	Low risk (Lo)	Medium risk (Me)	High risk (Hi)
Have you ever lost 25% or more of your investment within a year?	No	No	Yes
What would you do if your portfolio loses 20% in value within one year?	Sell immediately	Do nothing	Buy more assets

Table 6. Analyzed Robo-Advisors

RA of category A (Portfolio structure & used products)	Country	RA of category B (Portfolio structure only)	Country	RA of category C (No portfolio structure)	Country
Acorns	USA	Bevestor	DEU	Ameritrade TD	USA
Ally Invest Managed Portfolios	USA	Comdirect - cominvest	DEU	Axos	USA
Asset-Builder	USA	Deutsche Bank - ROBIN	DEU	Hedgeable	USA
Autowealth	SGP	ELVIA e-invest	CHE	Minveo	DEU
Baloise Monviso	DEU	E*Trade Core Portfolios	USA	Personal Capital	USA
Easyfolio	DEU	FutureAdvisor	USA	Quirion	DEU
E-Base	DEU	Investify	DEU	Smavesto	DEU
Ellevest	USA	Moneyfarm	GBR	Tradeking (Merged with Ally)	USA
Evestor	GBR	Schwab Intelligent Portfolios	USA	truevest	DEU
Fairr.de	DEU	SigFig	USA	Vaamo (Merged with Moneyfarm)	GBR
Fintego	DEU	Visualvest	DEU	Wealthfront	USA
Ginmon	DEU	Warburg Navigator	DEU	WiseBanyan (Merged with Axos)	USA
Growney	DEU	Whitebox	DEU	WMD Capital	DEU
JPMorgan Chase	USA			Zeedin Hauck-Aufhäuser	DEU
LIQID	DEU				
LIQID: *Sustainability*	DEU				
Merrill Guided Investing	USA				
Morgan Stanley	USA				
Morgan Stanley: *Climate*	USA				
Morgan Stanley: *Gender Diversity*	USA				

(*continued*)

Table 6. (*continued*)

RA of category A (Portfolio structure & used products)	Country	RA of category B (Portfolio structure only)	Country	RA of category C (No portfolio structure)	Country
Nutmeg	GBR				
Scalable Capital	DEU				
Wüstenrot	DEU				

References

1. Alt, R., Puschmann, T.: Digitalisierung der Finanzindustrie Grundlagen der Fintech-Evolution. Springer Gabler, Heidelberg (2016). https://doi.org/10.1007/978-3-662-50542-7
2. Jung, D., Dorner, V., Glaser, F., Morana, S.: Robo-advisory - digitalization and automation of financial advisory. Bus. Inf. Syst. Eng. **60**, 81–86 (2018)
3. Gomber, P., Kauffman, R.J., Parker, C., Weber, B.W.: On the fintech revolution: interpreting the forces of innovation, disruption, and transformation in financial services. J Manage. Inform. Syst. **35**, 220–265 (2018)
4. Sironi, P.: FinTech innovation - from robo-advisory to goal based investing and gamification. Wiley, Chichester (2016)
5. D'Acunto, F., Prabhala, N., Rossi, A.G.: The promises and pitfalls of robo-advising. Rev. Financ. Stud. **32**, 1983–2020 (2019)
6. Jung, D., Weinhardt, C.: Robo-advisors and financial decision inertia: how choice architecture helps to reduce inertia in financial planning tools. Presented at the Thirty Ninth International Conference on Information Systems (ICIS) (2018)
7. Robo-Advisor. https://www.robo-advisor.de/. Accessed 25 July 2020
8. Brokervergleich. https://www.brokervergleich.de/. Accessed 25 July 2020
9. Cambridge English Dictionary. https://dictionary.cambridge.org/us/dictionary/english/robot&.../adviser. Accessed 25 July 2020
10. Beketov, M., Lehmann, K., Wittke, M.: Robo advisors: quantitative methods inside the robots. J. Asset Manage. **19**, 363–370 (2018)
11. Faloon, M., Scherer, B.: Individualization of robo-advice. J. Wealth Manage. **20**, 30–36 (2017)
12. Nussbaumer, P., Matter, I., Reto à Porta, G., Schwabe, G.: Design für Kostentransparenz in Anlageberatungsgesprächen. Wirtschaftsinformatik. **54**, 335–350 (2012)
13. Tertilt, M., Scholz, P.: To advise, or not to advise—how robo-advisors evaluate the risk preferences of private investors. J. Wealth Manage. **21**, 80–84 (2018)
14. Jung, D., Glaser, F., Köpplin, W.: Robo-advisory: opportunities and risks for the future of financial advisory. In: Advances in Consulting Research, pp. 405–427 (2019)
15. Reher, M., Sun, C.: Automated financial management: diversification and account size flexibility. J. Invest. Manage. **31**(2), 1–13 (2019)
16. Scherer, B.: Algorithmic portfolio choice: lessons from panel survey data. Fin. Markets. Portfolio Mgmt. **31**(1), 49–67 (2017). https://doi.org/10.1007/s11408-016-0282-8

17. Huxley, S.J., Kim, J.Y.: The Short-Term Nature of Robo Portfolios. Advisor Perspectives (2016)
18. Amadi, F.Y., Amadi, C.W.: Investment Horizon and the Choice of Mutual Fund. IJBM **14**, 76 (2019)
19. Warren, G.: Long-term investing: what determines investment horizon? CIFR Paper No. 39, 1–39 (2014)
20. Steiner, M., Bruns, C., Stöckl, S.: Wertpapiermanagement: Professionelle Wertpapieranalyse und Portfoliostrukturierung. Schäffer-Poeschel, Stuttgart (2017)
21. Mondello, E.: Finance: Angewandte Grundlagen. Gabler Verlag, Wiesbaden (2018). https://doi.org/10.1007/978-3-658-21579-8
22. Berk, J., Demarzo, P.: Corporate Finance. Global Edition, Pearson (2019)
23. Markowitz, H.: Portfolio selection. J. Finan. **7**, 77–91 (1952)
24. Sharpe, W.F.: Mutual fund performance. J. Bus. **39**, 119–138 (1966)
25. Sharpe, W.F.: The sharpe ratio. J. Portfolio Manage. **21**, 49–58 (1994)
26. Janssen, J., Laatz, W.: Statistische Datenanalyse mit SPSS: Eine anwendungsorientierte Einführung in das Basissystem und das Modul Exakte Tests. Gabler Verlag (2017)
27. Field, A.: Discovering Statistics Using IBM SPSS Statistics. SAGE Publications Ltd., Thousand Oaks (2017)
28. Statista: Savings rate of households in selected countries worldwide from 2010 to 2018. https://www.statista.com/statistics/246296/savings-rate-in-percent-of-disposable-income-worldwide/. Accessed 25 July 2020
29. Statista: Robo-Advisors - United States. https://www.statista.com/outlook/337/109/robo-advisors/united-states?currency=eur. Accessed 25 July 2020
30. Statista: Robo-Advisors – Germany. https://www.statista.com/outlook/337/137/robo-advisors/germany?currency=eur. Accessed 25 July 2020
31. Statista: Net private financial assets per capita by country 2018. https://www.statista.com/statistics/329074/net-private-financial-assets-per-capita-worldwide/. Accessed 25 July 2020

Invited Talk

The Financial Viability of eHealth and mHealth

Bernard Le Moullec[1], Yan Hanrunyu[2], and Pradeep Kumar Ray[2,3(✉)]

[1] Data for Good, Paris, France
[2] Center for Entrepreneurship, University of Michigan-Shanghai Jiao Tong University Joint Institute, Shanghai, China
p.ray@unsw.edu.au
[3] School of Public Health and Community Medicine, UNSW, Sydney, Australia

Keywords: Economic evaluation · eHealth · mHealth

1 Introduction

While the demand for health services in increasing drastically, the scope for the increase in supply is limited due to cost and other constraints. The WHO has announced that e-Health (healthcare using Information and Communication Technologies) might have the potential to support healthcare providers in meeting the growing demand without sacrificing quality. E-Health also includes mHealth (healthcare using mobile phones). The recent COVID-19 has caused an explosion in e-Health (also called telehealth) use worldwide to prevent the infections through face-to-face contact between patients and health professionals (doctors, nurses etc.), many of whom have reportedly died of COVID-19 while treating patients.

Although many trials have shown the technological feasibility of e-Health, it is important to establish the economic viability of e-Health. While most publications restrict to mere cost comparison, it is important to consider cost utility analysis or try to assess the level of use eHealth services that need to be at least as cost efficient as current health care practices. This paper uses the Cost Effectiveness Analysis (CEA) approach illustrate the use of eHealth in one type of healthcare (Occupation Therapy-OT) vis-a-vis a traditional paper based approach.

2 eHealth/mHealth Costs

Costs are divided into three general classes: direct, indirect and intangible costs. Examples of direct cost are initial capital and investment costs (medical, video and telecommunication equipment, software), continuing operating costs of ICT (e.g. user charge of equipment for rental and maintenance, costs of communication), wages of doctors and other staff, costs of patients etc. Indirect costs include costs of caregivers, administrative costs, education and training for the technology and update skills, changes in workflow, processes and organization (project and change management). Intangible costs

B. Clapham and J.-A. Koch (Eds.): FinanceCom 2020, LNBIP 401, pp. 111–116, 2020.
https://doi.org/10.1007/978-3-030-64466-6_7

include costs of software quality, productivity of people caused by the work environment, processes etc.

The most commonly applied economic evaluation techniques are cost-minimization analysis, cost-benefit analysis, cost-effectiveness analysis and cost-utility analysis [1].

Cost-effectiveness analysis (CEA) compares the relative expenditure (costs) and outcomes (effects) of several programs. As far as healthcare is concerned, the cost-effectiveness of a given health program is the ratio of the cost of the intervention to a measure of its effect. Here, the word "cost" refers to the resource used for the intervention, usually expressed in monetary terms. The measure of effects, expressed in "natural" units, changes depending on the intervention being considered. Cost-effectiveness is expressed as an "Incremental Cost Effectiveness Ratio" (ICER), which can be summed up by "the ratio of change in costs to the change in effects". A particular case of CEA is the CUA (Cost-Utility Analysis), where the effects are expressed in terms of years of full health lived, using metrics such as QALYs (Quality-Adjusted Life Years) or DALYS (Disability-Adjusted Life Years). CEA can be used when the outcomes of the compared technologies can be expressed in the same units. ICER is calculated using the following formula:

$$ICER = (C1-C2)/(E1-E2)$$

Where $C1$ = the cost of the new intervention, $C2$ = the cost of the comparator, $E1$ = the effect of the new intervention, and $E2$ = the effect of the comparator. With CEA, analysts often use a decision analytic approach (i.e. a complex mathematical modeling technique) that builds upon the long term costs and effectiveness. This paper will derive a modified cots model for mHealth in Sect. 3 (Occupation Therapy case study).

3 Occupation Therapy Case Study

This scenario demonstrates an example application of mobile health in the daily work of an Occupational Therapist (i.e. "a profession concerned with promoting health and well-being through occupation, whose primary goal of occupational therapy is to enable people to participate in the activities of everyday life" according to the World Federation of Occupational Therapists).

Many of the problems and risks that show up during the analysis of the occupational therapist assessment process (e.g. illegible handwriting, own abbreviations, waste of time, difficult collaborative work due to paperwork, security and privacy) could be addressed by integrating a mobile health enabled device into the assessment and communication process. The health record of the patient would be stored electronically in an Electronic Health Record (EHR) system in the hospital.

1. Review of the patient's health record
2. Creation of a case report
3. Interview of the patient according to the case report card
4. Assessment of the patient

The next Section compares the traditional paper based model with an eHealth system based on EHR described above. The direct costing or Marginal is a costing model that includes only the variable costs (direct materials, direct labour, etc.). Here we focus on direct costs only.

3.1   Actual OT Paper-Based Assessment Process

a. Patient health record may not be available. In the worst case, a new health record might have to be created from scratch, with the risk of incomplete information and/or a redundancy with the original health record, creating the risk of additional confusion.
b. A case report card is filled with information from the health record may involve the duplication, resulting in a loss of time and productivity.
c. Feedback and assessment of the patient: substantial time and transportation costs are involved if the OT deals with patients coming from a remote village or rural area.
d. The collaborative work is complicated by the fact that all information and results are stored on paper. The cost of physical storage is added to the opportunity cost of medical staff treating other patients.

3.2   Proposed mHealth-Based OT Assessment Process

1. The OT reviews the health record of the patient. The OT accesses the EHR-System via the mobile EHR device and looks up the health record via the patient's name.
2. To create a new case report to assess his/her patient, the OT selects the corresponding menu button. The time for the initial completion of the case report header and personal details section is greatly reduced.
3. The OT does not waste any time anymore for copying information from the health record into the case report. Using the mobile EHR-System, the OT does not use his/her handwriting anymore.
4. Appointments are handled electronically, also when the OT works in a rural hospital and needs consultation of a specialist located several hundreds of kilometers away. The mHealth solution virtually eliminates every transportation need and speeds up the whole consultation process, generating time savings and decreasing travel costs.
5. The case report can be accessed and shared by several people at the same time.

3.3   Tentative Direct Cost Analysis Model

We focus on the two main direct cost components: direct labour and direct supplies (subdivided into direct fixed supplies, direct fixed labour, direct variable supplies, and direct variable labour). The following analysis will be carried on the basis of profitability.

In order to make the comparison, 2 groups should be constituted: the control/usual care group (whose total cost of care is represented by the output of the function F_p) and the intervention/treatment group (whose total cost of care is represented by the output of the function F_m). In our case, the main variable is the number of OT consultations (i.e. OT cases treated) done with/without the support of mHealth technologies. We could call it "t". Other parameters are listed in Table 1.

Table 1. Parameter definition (TBD stands for "To Be Determined")

F_p	Global sum of all the direct costs related to the **paper-based OT** assessment workflow
F_m	Global sum of all the direct costs related to the **mHealth OT** assessment workflow
(TBD) a	Travelling costs for OT (time and distance related)
(TBD) b_m	Fixed capital start-up costs (nil in the case no mHealth technology is implemented)
(TBD) c_p/c_m	The associated workload for OT and doctors in FTE (the workload associated with the recurrent losses of time with the old paper system, but also the initial workload associated with the mHealth training) per consultation. The number of hours worked is used as a quantity index for labour
(TBD) d	The wage costs of both OT and doctors per FTE
(TBD) e_m/e_p	Costs of communication per consultation, should it be electronic or through paper and face-to-face interview, both between the OT and the patient and then the OT and the doctor
(TBD) f_p/f_m	Costs of data storage per consultation, should it be on hard copy or digital (mHealth solution or paper-based solution)
(TBD) g_p/g_m	Average number of days a consulted patient stays in the hospital, depending on which solution is used (mHealth solution or paper-based solution)
(TBD) h	Costs of admission/staying of a consulted patient into the hospital (per day) (including medicine, food, medical supplies, administrative costs, energy)
(TBD) i_m	Decrease in medical expenses for patients (per consultation) due to increased efficiency of healthcare

The cost function is a function whose parameters are the given output level (in our case health) and the given factor prices (listed below). The purpose of microeconomic analysis is to determine the combination of factor prices (for instance, the process/technological model) that minimizes healthcare cost while maintaining the same level of output. Here our approach is slightly different: we try to assess which combination of factor prices (i.e. which OT process) is the least costly while assuming the mHealth "combination" is as least as effective as the paper-based one.

For simplicity reasons, we will assume our direct cost function is linear, i.e. a first-degree polynomial function of one variable, in our case, the number of uses: since the number of uses is necessarily discrete, the cost function could have been a progression or a sequence, however we chose to stick to the function model in order to allow our main variable to be replaced by any continuum, such as time (relying on the trivial assumption that the number of consultations done grows as time goes on). Therefore, let's assume that t belongs to $\mathbb{R}_+$. Assuming our direct cost function is linear will make the job easier (basically saying that "Total Cost = Fixed Costs + Variable Costs * The Variable Itself), but will not reflect faithfully benefits such as experience gains and productivity improvements along the studied time period (this is why we dwell on another potential method later on). For any t in $\mathbb{R}_+$, assuming the cost function is linear, we shall define $F_m(t)$ and $F_p(t)$ as the following:

$$F_m(t) = b_m + t * (a + c_m * d + e_m + f_m + g_m * h - i_m)$$

and

$$F_p(t) = t * (a + c_p * d + e_p + f_p + g_p * h)$$

Thus

$$F_m(t) - F_p(t) = b_m + t * (d * [c_m - c_p] + [e_m - e_p] + [f_m - f_p] + h * ([g_m - g_p] - i_m)$$

As a sequence,

$$F_m(t) - F_p(t) >= 0 \Leftrightarrow t \le - b_m / (d * [c_m - c_p] + [e_m - e_p] + [f_m - f_p] + h * ([g_m - g_p] - i_m)$$

We infer from the equation above that our analysis will have to focus on the fixed startup costs for the mHealth solution and the difference in communication costs, workload, data storage costs, number of days a consulted patient stays in hospital, let alone the decrease in medical expenses for patients (per consultation) due to increased efficiency of healthcare (im). Once these formulas have been used, we could also use a linear regression technique to assess the correlation between the use of mHealth and the cost of OT interventions on a given period of time.

From the tradition of using cost functions to explain observed variations in unit costs, we could estimate a long-run cost-function using an OLS (ordinary Least Squares regression analysis). The regression equation deals with the following variables: the unknown parameters denoted as β (some of the costs listed above, for instance; this set of unknown parameters may be a scalar or a vector), the independent variable, t, and the dependent variable, C (the unit cost for OT intervention). The model for estimation (one-way fixed effect model) would look like this (respectively without and with mHealth):

$$C = F_m(t, \beta)/t \quad and \quad C = F_p(t, \beta)/t$$

Since this analysis is oriented straight-forward to direct cost assessment, it doesn't account for non-monetary benefits achieved throughout the implementation of mHealth in the occupational therapy workflow, such as less anxiety in day-to-day life, enhanced consciousness towards health, thus ignoring some of the multidimensional aspects of the output provided by hospitals.

4 Conclusions

The economic viability is an extremely important aspect if eHealth and mHealth have to be adopted globally including developed and developing countries. This paper has discussed the cost benefit of mHealth compared to traditional paper based (and face-to-face consultation) systems in the context of Occupation therapy, a major healthcare method for rehabilitation from long-term illnesses, such as a stroke or mental illness. A mathematical model has been derived to compare eHealth/mHealth costs with traditional face-to-face, paper based method. The application shows cost savings in eHealth/mHealth but the exact amount depends on the nature of healthcare, country of application etc. However, eHealth/mHealth has become widely accepted during the COVID-19 all over the world, to reduce coronavirus infections between patients and health professionals.

Reference

1. Preedy, V.R., Watson R.R. (eds.): Handbook of Disease Burdens and Quality of Life Measures. Springer, New York. https://doi.org/10.1007/978-0-387-78665-0_5565

Author Index

Printed in the United States
By Bookmasters